ENERGY POLICIES, POLITICS AND PRICES

BUILDINGS, RESOURCE CONSUMPTION AND OCCUPANT BEHAVIOR

STRATEGIES FOR CHANGE AND THE ROLE OF SUBMETERING

ENERGY POLICIES, POLITICS AND PRICES

Additional books in this series can be found on Nova's website under the Series tab.

Additional e-books in this series can be found on Nova's website under the e-book tab.

ENERGY POLICIES, POLITICS AND PRICES

BUILDINGS, RESOURCE CONSUMPTION AND OCCUPANT BEHAVIOR

STRATEGIES FOR CHANGE AND THE ROLE OF SUBMETERING

JOANNA CRUZ
EDITOR

nova publishers
New York

NOTICE TO THE READER

Library of Congress Cataloging-in-Publication Data

ISBN: 978-1-63482-767-6

Published by Nova Science Publishers, Inc. † New York

CONTENTS

PREFACE

The people who use Federal buildings — Federal employees, operations and maintenance staff, and the general public — can significantly impact a building's environmental performance and the consumption of energy, water, and materials. Many factors influence building occupants' use of resources (use behaviors) including work process requirements, ability to fulfill agency missions, new and possibly unfamiliar high-efficiency/high-performance building technologies; a lack of understanding, education, and training; inaccessible information or ineffective feedback mechanisms; and cultural norms and institutional rules and requirements, among others. While many strategies have been used to introduce new occupant use behaviors that promote sustainability and reduced resource consumption, few have been verified in the scientific literature or have properly documented case study results. This book documents validated strategies that have been shown to encourage new use behaviors that can result in significant, persistent, and measureable reductions in resource consumption. This book also presents guidance and recommendations on the benefits and complexities in the use of submetering technologies for new and existing buildings.

Chapter 1 – The construction and buildings sector consumes enormous amounts of energy, clean water, and materials. Buildings use about one-third of the world's energy. In the United States today, the buildings sector accounts for 40 percent of the primary energy use, compared to 32 percent for the industrial sector and 28 percent for the transportation sector. The use of electric power and heat in the buildings sector also accounts for about 40 percent of U.S. greenhouse gas (GHG) emissions. Advances in water- and energy efficient technologies hold the promise to significantly reduce resource consumption in buildings. However, history repeatedly has shown a gap between potential and actual energy, water, and material savings in buildings

— a gap often associated with the behavior of building operators and occupants. Further, many proposed strategies for changing behavior to reduce resource consumption are unverified and their outcomes are undocumented. For the Nation to significantly decrease its energy, water, and material consumption in buildings, a combination of new technologies and proven strategies for changing occupant and organizational use practices is required. Responding to this need, this paper offers strategies and methods for change at individual, organizational, and institutional levels of action to reduce energy, water and materials consumption.

Chapter 2 – This report presents guidance and recommendations on the benefits and complexities in the use of submetering technologies for new and existing buildings. Proven demonstrations of new building technologies are essential to high-performance and sustainable building designs and operational practices that positively impact the effective stewardship of our Nation's resources. A promising avenue for advancement of these goals involves the deployment of advanced building instrumentation and submetering for real-time measurement of energy and water usage. To this end, this report documents the current state of submetering, relevant case studies and preliminary findings relating to submetering system costs and return on investment. It provides references to relevant publications for continual and accurate measurement of resource consumption, focused design and retrofit strategies, improved building management procedures, and changes in human behaviors that lead to significant reductions in energy and water usage.

In: Buildings, Resource Consumption … ISBN: 978-1-63482-767-6
Editor: Joanna Cruz

Chapter 1

BEHAVIORAL CHANGE AND BUILDING PERFORMANCE: STRATEGIES FOR SIGNIFICANT, PERSISTENT, AND MEASURABLE INSTITUTIONAL CHANGE*

Amy K. Wolfe, Elizabeth L. Malone, Judith Heerwagen and Jerome Dion

SUMMARY

The people who use Federal buildings — Federal employees, operations and maintenance staff, and the general public — can significantly impact a building's environmental performance and the consumption of energy, water, and materials. Many factors influence building occupants' use of resources (use behaviors) including work process requirements, ability to fulfill agency missions, new and possibly unfamiliar high-efficiency/high-performance building technologies; a lack of understanding, education, and training; inaccessible information or ineffective feedback mechanisms; and cultural norms and institutional rules and requirements, among others. While many strategies have been used to introduce new occupant use behaviors that promote sustainability

* This is an edited, reformatted and augmented version of a report, PNNL-23264, prepared by Pacific Northwest National Laboratory under contract with the U.S. Department of Energy, April 2014.

and reduced resource consumption, few have been verified in the scientific literature or have properly documented case study results.

This paper documents validated strategies that have been shown to encourage new use behaviors that can result in significant, persistent, and measureable reductions in resource consumption. From the peer-reviewed literature, the paper identifies relevant strategies for Federal facilities and commercial buildings that focus on the individual, groups of individuals (e.g., work groups), and institutions — their policies, requirements, and culture. The paper documents methods with evidence of success in changing use behaviors and enabling occupants to effectively interact with new technologies/designs. It also provides a case study of the strategies used at a Federal facility — Fort Carson, Colorado. The paper documents gaps in the current literature and approaches, and provides topics for future research.

Institutional context, composed of three major elements (rules, roles, and tools), frames the strategies and actions that drive significant and enduring reductions in energy, water, and material use. New approaches must use evidence-based principles and apply proven behavioral change strategies that align with the behaviors and cultures at multiple levels within the organization and that are best suited for achieving specific energy- or sustainability-related goals. The key strategies for promoting behavioral change evident in peer-reviewed literature include

- Providing Information and Outreach
- Giving Feedback
- Persuading
- Offering Rewards and Incentives
- Changing Defaults, Nudging Choice, and Enabling Adaptive Comfort
- Gaining Commitment
- Tapping into Social Norms
- Social Marketing in Communities
- Responding to an Urgent Situation.

Initiating change is different from maintaining or improving upon those changes over time. Ensuring sustained change requires an adaptive approach that (1) leverages successes, (2) updates and embeds formal and informal organizational rules to support sustainable resource use, and (3) strategically promotes change in organizational culture driven by leaders and others playing key roles in the organization.

This paper provides the evidentiary and conceptual background for selecting proven context- and goal-specific behavioral change strategies, and highlights potential opportunities for Federal agencies and the commercial building sector. The discussion employs a "systems perspective" of occupant use behaviors to drive broad organizational transformation in support of sustainability and improved building performance. The objective is to elucidate potential strategies to achieve improved building performance within the context of the organization and its culture — behavior and behavior change aim to enable these outcomes and are not ends in themselves nor mandated by Federal requirements.

The paper makes the following high-level recommendations for implementing an action-based framework for changing individual and organizational use behaviors and achieving organizational sustainability and building performance goals:

1) Adopt a systems perspective to assess, design, implement, and evaluate sustainability and resource use strategies and programs as integrated technological and use activities.
2) Employ multiple methods and strategies that are supported by peer-reviewed scientific research as effective and aligned with specific sustainability goals and with mission fulfillment.
3) Promote an enduring institutional transformation by evaluating outcomes to determine what strategies work best to promote and reinforce a culture of sustainability, what strategies are not effective and should be discontinued, and what strategies should be revised.
4) Sponsor social science research that will build the evidence base for behavioral and institutional strategies aimed at saving energy, water, and materials, and at operating buildings sustainably over the long term.

Unless new technologies, occupant use behaviors, and institutional rules and requirements are integrated to complement new stringent Federal building performance requirements, suboptimal outcomes will limit Federal agencies' abilities to achieve their sustainability and building performance goals and objectives now and in the future.

ACRONYMS AND ABBREVIATIONS

BEM	Building Energy Monitor
BTRD Subcommittee	Buildings Technology Research and Development Subcommittee
CBSM	Community Based Social Marketing
CFL	Compact Fluorescent Lamp
DOD	U.S. Department of Defense
DOE	U.S. Department of Energy

GHG	Greenhouse Gas
NSTC	U.S. National Science and Technology Council
O&M	Operations and Maintenance
OSTP	U.S. Office of Science and Technology Policy

PREFACE

The construction and buildings sector consumes enormous amounts of energy, clean water, and materials. Buildings use about one-third of the world's energy. In the United States today, the buildings sector accounts for 40 percent of the primary energy use, compared to 32 percent for the industrial sector and 28 percent for the transportation sector. The use of electric power and heat in the buildings sector also accounts for about 40 percent of U.S. greenhouse gas (GHG) emissions. Advances in water- and energy-efficient technologies hold the promise to significantly reduce resource consumption in buildings.

However, history repeatedly has shown a gap between potential and actual energy, water, and material savings in buildings — a gap often associated with the behavior of building operators and occupants. Further, many proposed strategies for changing behavior to reduce resource consumption are unverified and their outcomes are undocumented. For the Nation to significantly decrease its energy, water, and material consumption in buildings, a combination of new technologies and proven strategies for changing occupant and organizational use practices is required.

Responding to this need, this paper offers strategies and methods for change at individual, organizational, and institutional levels of action to reduce energy, water and materials consumption. The paper communicates some of the best practices examined by the Buildings Technology Research and Development (BTRD) Subcommittee of the Committee on Technology (COT), National Science and Technology Council (NSTC) and presents a framework for integrating human behavior, organizational practices, and building operations to enhance overall building performance. The paper is intended for Federal executives, program managers, sustainability officers, and procurement and human resource specialists with responsibility to achieve Federal building performance and sustainability requirements. The framework applies in principle to Federal and commercial office, laboratory, and other nonresidential buildings designed for occupant use.

1.0. INTRODUCTION

Architects and engineers design and retrofit Federal buildings to achieve increasingly aggressive environmental goals and mandates using a broad range of advanced technologies and design strategies. These technologies and designs, for new and retrofitted high-performance buildings, may perform well in theory — based upon analytical building models and simulations. Once the building is completed and occupants move in, however, the actual performance often runs contrary to expectations. While some high-performance buildings achieve their environmental performance goals, many do not (Turner and Frankel 2008, Cole 2011). History repeatedly has shown a gap between potential and actual energy savings, a gap often associated with the behavior of people who manage and use buildings and building systems.

Increasing evidence (e.g., Stern 2002, Janda 2014) suggests that the everyday behaviors of building occupants and standard operating procedures of organizations are a much-overlooked factor in achieving building performance goals and the conservation of resources (energy, water, materials). Included in these everyday behaviors are standard (and substandard) building operating procedures, occupant use (and misuse) of building resources and technologies, and outdated organizational practices and cultural norms. For office buildings, the number of miscellaneous devices plugged into electrical outlets continues to rise and represents the single fastest-growing category for electrical consumption in buildings. When combined with an ever-increasing reliance on other energy-intensive technologies and equipment (e.g., computer servers, high-quality printers) to carry out agency mission requirements, occupant use behaviors make conserving resources and improving building performance an increasingly important and challenging endeavor. Janda (2014) suggests that the "social potential" of energy efficiency in workplaces is as or more important than the "technical potential" usually measured – but that an approach integrating the two would have the largest impact (Shui 2012, Moezzi and Janda 2013).

This paper, commissioned by the former Behavioral Task Group for the Buildings Technology Research and Development Subcommittee of the National Science and Technology Council, provides a framework for integrating human behavior, organizational practices, and building operations to enhance overall building performance. It offers strategies and methods for change at the individual, organizational, and institutional levels. Section 2 reviews the importance of a focus on human-centered change, followed by an overview of ways to think about energy use and behavioral change in buildings (Section 3). Section 4 uses a construct of "rules, roles, and tools" as to guide the assessment of institutional context and as a basis for selecting strategies that are effective for achieving and maintaining behavioral and institutional change. Sections 5 and 6 take a closer look at successful approaches for changing use behaviors, and Section 7 presents a case study of behavioral change in a Federal facility. The paper's conclusions (Sections 8 and 9), make recommendations for an action-oriented framework for improving use behavior in buildings, and for an R&D agenda aimed at deepening the understanding of sustainability-oriented behavior and behavioral change to make programs for significant, persistent change more effective.

2.0. Why Behavior Matters

Understanding behavior is crucial in determining the extent to which performance goals embedded in building design and technologies actually are achieved. Building occupants are seldom trained or well informed about building systems operation and how their behaviors, individually and collectively, affect a building's environmental performance. Building operations and maintenance (O&M) staff may not have adequate training to properly operate new, high-performance, energy-efficient building systems. Furthermore, interventions designed to improve building performance are often implemented piecemeal, focusing on a single component and target audience at a time rather than viewing the building and its occupants[1] as an integrated whole (i.e., a systems view) with varying needs and space types, technologies, and workplace responsibilities and practices. Building technologies designed to reduce resource consumption must therefore be complemented by appropriate use behaviors by O&M staff and occupants. Identifying effective combinations of technologies and behaviors can reduce resource consumption resulting in lower facility operational costs.

[1] "Occupants" include organizations and all individuals associated with the organization. Some organizational actors may not actually work within a particular building, yet have great influence over energy-use behaviors within the building.

Theory — supported by scientific research — indicates that behavior can play a substantial role in reducing greenhouse gas (GHG) emissions and energy use (Dietz et al. 2009). However, despite recent interest in and efforts to realize this potential, results have been both disappointing (i.e., not meeting aspirational goals) and lacking in the kinds of evidence that could guide future efforts (i.e., without the basis for confident replication) (Lenoir et al. 2011, Hauge et al. 2010).[2] In the absence of coherent and well-documented approaches to changing behavior and organizational change management, there has been a burgeoning interest in exploring new approaches. A negative consequence of such activity is that untested and potentially ineffective strategies may be adopted, resulting in suboptimal outcomes that can discourage building owners, managers, and occupants from embracing future sustainability initiatives.

2.1. User-Building Interface

The importance of behavioral change is evident in that building systems using new high-performance technologies often fail to meet their "as designed" performance expectations (Turner and Frankel 2008). At least part of this performance gap is likely attributable to a mismatch between the building's design elements and the users' needs — the user-building interface — as well as to occupants' unforeseen uses of the building systems and technologies. As noted by Turner and Frankel, the highest energy use buildings are "driven by requirements and process loads of occupant activities as opposed to basic building systems." For Federal agencies these "occupant activities" can be characterized as mission-relevant.

At the present time, designers and operators often do not adequately address the user-building interface. This omission has significant consequences for how buildings are designed and equipped, how well buildings achieve sustainability goals, and how building systems can affect the work environment and occupant comfort levels. For energy performance goals, the problem is exacerbated by the fact that "energy is not consumed directly, but rather as a means to achieve specific goals" which include comfortable temperatures, sufficient light, and the ability to use computers and other electronic equipment necessary for work (Aune et al. 2009). Thus, ***the***

[2] The disappointing results for energy efficiency initiatives are a specific example of the general disappointing results of corporate change initiatives (see Aiken and Keller 2009).

building occupant's perspective on energy consumption is not about energy per se, but the capabilities it provides for achieving job and mission goals. Deep, sustained changes in occupant use behavior require a systematic, well-designed approach — more than the simple "competitions," lobby energy displays, or desktop dashboards currently in vogue. As this paper demonstrates, persistent and transformative change is built upon a strategic approach involving multiple and continuing change efforts throughout the organization and a rethinking of the user-building interface (Moezzi and Janda 2013, Cole 2011).

2.2. Individual Behaviors and Norms

Human behavior is an essential and frequently overlooked element in achieving aggressive Federal building performance goals, especially for reducing consumption of energy, water, and materials. Building technologies are increasingly effective at reducing resource consumption, but occupant routines, habits, and choices can significantly impact the building's overall performance. These behaviors include many habitual actions (e.g., turning off equipment when not being used, routine building system O&M procedures) as well as complex behaviors, such as procurement or contracting procedures, required to implement policies that affect high-impact outcomes (mission, energy use, and so on). Some behaviors may be episodic, e.g., responding to emergency conditions such as rolling brownouts or water shortages. Other behaviors may involve a suite of actions that drive sustainability across the organization through changes in roles, responsibilities, and ways of thinking. Deep, sustained change at the individual level is more difficult to achieve and requires a programmatic focus and significant effort for perhaps several years until new norms become embedded as "how things are done" in the organization.

2.3. Sustainability Efforts in Federal Buildings

In the Federal sector, new governmental rules contained in Presidential Executive Orders, Office of Management and Budget (OMB) Circulars, and statutory requirements are driving Federal sustainability efforts. Of these, the Energy Independence and Security Act of 2007 (EISA 2007) requires agencies to establish and report on increasingly aggressive sustainability and performance goals on an annual basis. To meet these goals, institutional procedures governing internal Federal agency operations, mission requirements, personnel roles and responsibilities, contracts, and procurements must now align with these new directives and annual goals. Moreover, institutional change must be aligned with the desired behavioral change throughout the workplace. Without these alignments, suboptimal outcomes will occur and hinder Federal agencies' abilities to achieve their mandated sustainability goals and objectives.

Federal agencies are now implementing strategies for achieving far-reaching sustainability goals, including increasing energy and water efficiency; reducing GHG emissions, waste, and pollution; and more effectively managing information technology resources. The growing recognition in the Federal community is that achieving these goals will involve large-scale technological change and significant behavioral changes at the individual, organizational, and institutional levels. New strategies must be identified that effectively integrate these technology and behavioral elements. Such strategies will produce better results than either technological *or* behavioral strategies by themselves (Shui 2012), reorienting occupants in their various rules of operation, roles, technologies, and infrastructure to ensure a continuous process of becoming more sustainable. The desired outcome is transformational, persistent change manifested in significant progress toward established goals in all sustainability and building performance categories. Successes in the Federal sector can then be shared with other important building subsectors: commercial office and other nonresidential buildings.

3.0. Linking Behavioral and Organizational Change to Achieve Sustainability Goals

This paper provides an integrative approach linking traditional and emerging behavioral change methods at individual and group levels with methods and best practices for organizational change. Most behavioral change efforts focus on either individual or organizational change and omit the synergy that results from the combination of approaches. Given the breadth of behaviors and practices that need to change to achieve deep and lasting sustainability in an organization, narrowly focused, short-term efforts will not meet current or future sustainability goals and requirements. To begin this analysis, definitions are given that will be used throughout and provide context for the discussion, followed by sections on buildings as the locus of behavior and the dynamics of building occupants.

3.1. Definitions of Behavior, Behavior Change, and Building Operational Performance

> *Behavior* includes how people and organizations act, how they function (patterns of action and interaction over time, sometimes referred to as institutional behavior), and what choices they make — whether in response to internal/external stimuli, or in anticipation of future events/conditions, out of subconscious habit, or for no immediately apparent reason.

Responding to an internal stimulus of feeling too warm, for example, may lead to opening a window, turning a thermostat down, or using an electrical fan. Responding to an external stimulus would be closing the window blinds to avoid glare. External stimuli can also be non-physical, such as a new rule or procedure that directs people to respond in a new way (such as turning their computers off at night). In addition, individuals respond to non-energy, non-resource stimuli and situations in ways that have significant impacts on resource consumption. They use high-quality printers, copiers, and video conferencing equipment to increase their productivity as well as espresso nooks, refrigerators, and microwaves to create more pleasant work environments. At an organizational level, people act in response to their job responsibilities (e.g., procurement versus maintenance versus administrative), which, in turn, are affected by a wide variety of expectations and stimuli.

> *Behavior change* refers to alterations in patterns of action and choices.

Behavior change happens all the time — typically not as a result of deliberate planning or intentional initiatives but more often as unintended consequences of poor planning or through the introduction of new, untested procedures. At the same time, some behaviors endure, even when the original reason for them has disappeared ("because we've always done it that way"). Even when individuals, organizations, and social groups have good intentions to change their own or organizational behaviors, these intentions can go awry.

Importantly for those seeking to effect change, behaviors often occur in sequences or patterns that may need to be changed as a whole. Behavioral patterns at both the individual level (e.g., technology use) and the organizational and institutional levels (e.g., purchasing and infrastructure decisions) have significant impacts on building operational performance. Technologies in use, policies and procedures, and infrastructure (the building shell, interior walls and surfaces, layout of space, equipment, etc.) are the result of choices driven by individuals' behaviors; however, once in place, they also serve to constrain and channel behavior. Therefore, this study uses inclusive definitions of "behavior" and "behavior change" encompassing individuals (those with the authority and responsibility to plan or control and those who are supposed to follow directions), organizations, and institutions. Moreover, the adoption and use of technologies are inextricably intertwined with behaviors at all levels as in the common phrase, "You can't have one without the other."

> *Building operational performance* is the measured metrics of system efficiency, sustainability, productivity, environmental quality, and functionality.

Building operational performance is considered here in two contexts. The first emphasizes energy and water use, as addressed in Executive Order 13423 (Strengthening Federal Environmental, Energy, and Transportation Management, from 2007)[3] and expanded under Executive Order 13514 (Federal Leadership In Environmental, Energy, and Economic Performance, from 2009).[4] In this sense, the goals are to minimize energy and water use through conservation and efficiency, and to increase the use of renewable types of energy and products. The second relates to sustainability, as per Executive Order 13514. In this sense, building operational performance includes an emphasis on energy, GHG emissions, and water use efficiency, but adds other sustainability goals such as waste minimization, pollution reduction, and information technology management.

Given these definitions, it follows that behavioral change in buildings must be addressed at multiple levels within the context of organization. The behavior of individuals in the organizational context is governed by the formal (explicit) and informal (implicit) expectations, rules, and regulations of the employment contract. The success of organizations in meeting their goals depends on effective application of the knowledge, skills, and abilities of individuals in their roles. As organizations continually evolve to meet internal

[3] http://www.gpo.gov/fdsys/pkg/FR-2007-01-26/pdf/07-374.pdf
[4] http://www.whitehouse.gov/assets/documents/2009fedleader_eo_rel.pdf

challenges (e.g., expansion, contraction, reorganization, new products/services) as well as external challenges (e.g., regulation, legal requirements, societal imperatives, competition), strategies must adjust to address these changes.

When behavioral and institutional modifications produce persistent change, the tenor of the organization changes, enabling it to establish and meet more ambitious sustainability goals as time goes on. Studies of major firms that have succeeded in implementing behavioral change (e.g., Prindle 2010, Duhigg 2012) show that not only are employees enthusiastic about devising ways to save energy, but also that such efforts spur innovation in a broader range of improvements to business processes. In addition, agencies gain a deeper understanding of their own culture and effective pathways to cultural change (Haanaes et al. 2011).

Institutionalizing these types of changes can help agencies build upon and adjust their use behavior strategies, in contrast to launching a series of time-limited initiatives with high expectations but short-lived effectiveness.

3.2. Buildings As the Locus of Behavior

Behaviors relevant to buildings occur at many levels, from individual to institutional (see Figure 3.1). Inside a building there may be open spaces with cubicles and reception areas, a cafeteria, conference rooms, hallways, stairways, restrooms, and elevators. The uses of these spaces can range widely: work processes, product development, meetings, social gatherings, and housing energy/water/mechanical systems (including lighting, heating, cooling, and ventilation). Building spaces must accommodate equipment (copiers, printers, telephones, vending machines), staffing changes, food and drink consumption, restroom comfort stations, and getting around (in and out of the building, to and from interior spaces).

The intent behind designing and operating buildings, of course, is to support an agency's work and mission. Any environmental goals should aim to provide this support in a sustainable way. When Federal managers and staff talk about building performance, therefore, the fundamental purpose of the building must be constantly kept in mind, and ideally explicitly linked to actions taken.

The everyday uses of space and equipment affect a building's overall performance. Building system data can be used to detect performance changes in building systems, internal and external building environments, and the use of spaces and systems in the building. The data can identify use patterns of the occupants as well as interactions among building systems and between those systems and the environment (see also Appendix A).

Behaviors also affect a building's overall performance and the ability of its hardware and mechanical infrastructure to achieve resource and environmental objectives through the decisions people make and how they select, use, and maintain the building and its infrastructure (hence the gap between "as designed" and actual building performance (Turner and Frankel 2008)). Likewise, the office equipment, policies, and procedures that organizations use result from individuals' decisions, and require people to select, use, maintain, and interact in ways that serve their business needs or mission.

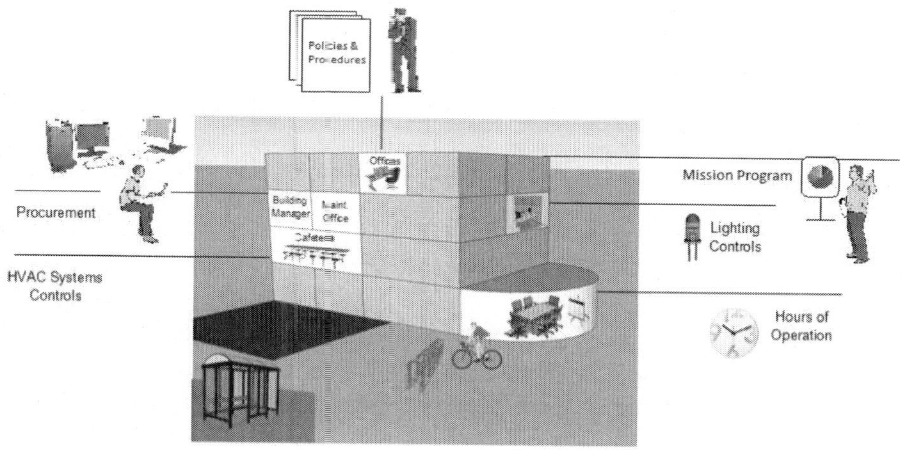

Figure 3.1. Building Resource Use is Influenced by Occupants, Systems, Infrastructure, and Organizational Rules.

3.3. The Dynamics of Occupants in Buildings

Buildings are complex systems, and so are the social and organizational activities within them. Building occupants, who are assigned building space or authorized to use its services, need tools and an environment to help them do their jobs—comfort and food services, adequate lighting, and ventilation — and to interact with other building occupants, and may have set work patterns or highly irregular and unpredictable schedules and duties.

An institution may establish rules for hours of operation as well as for conditions of vacancy (e.g., telecommuting arrangements, travel). Building owners and managers typically use infrastructure and equipment to provide adequate working conditions, amenities, and comfort. Procurement staff may acquire equipment and supplies (e.g., paper, toner, batteries) to support occupant needs and well-being. Decisions and actions may be based on formal directives or informal practices (e.g., if high-level managers complain about ambient temperature, energy or facility managers are likely to make adjustments, regardless of the energy/performance implications). And, the specifics of all of these elements may change over time.

Thus, sustainability and building performance objectives may influence, but not necessarily determine, what decisions are made or what actions are taken. This dynamic environment necessitates a flexible and evolving approach to aligning occupant use behaviors with building functions, services, and systems over time.

4.0. Rules, Roles, and Tools: Using Knowledge of Institutional Context as a Basis for Taking Action

The complex and dynamic world of human behavior and behavior change affects both the performance of and the demands on Federal and commercial office buildings. The three elements (rules, roles, and tools) outlined in this paper help to frame the evidence-based insights and actions that drive significant and enduring reductions in energy and resource use. These elements provide avenues for understanding the institutional context within which improved building performance is to be achieved (see Figure 4.1). Viewing organizations and building performance through the lens of rules, roles, and tools provides a basis for selecting potentially effective strategies to spur and maintain desired changes. From this standpoint, behavior and behavior change help to enable desired outcomes; they are neither ends in themselves nor mandated by Federal requirements.

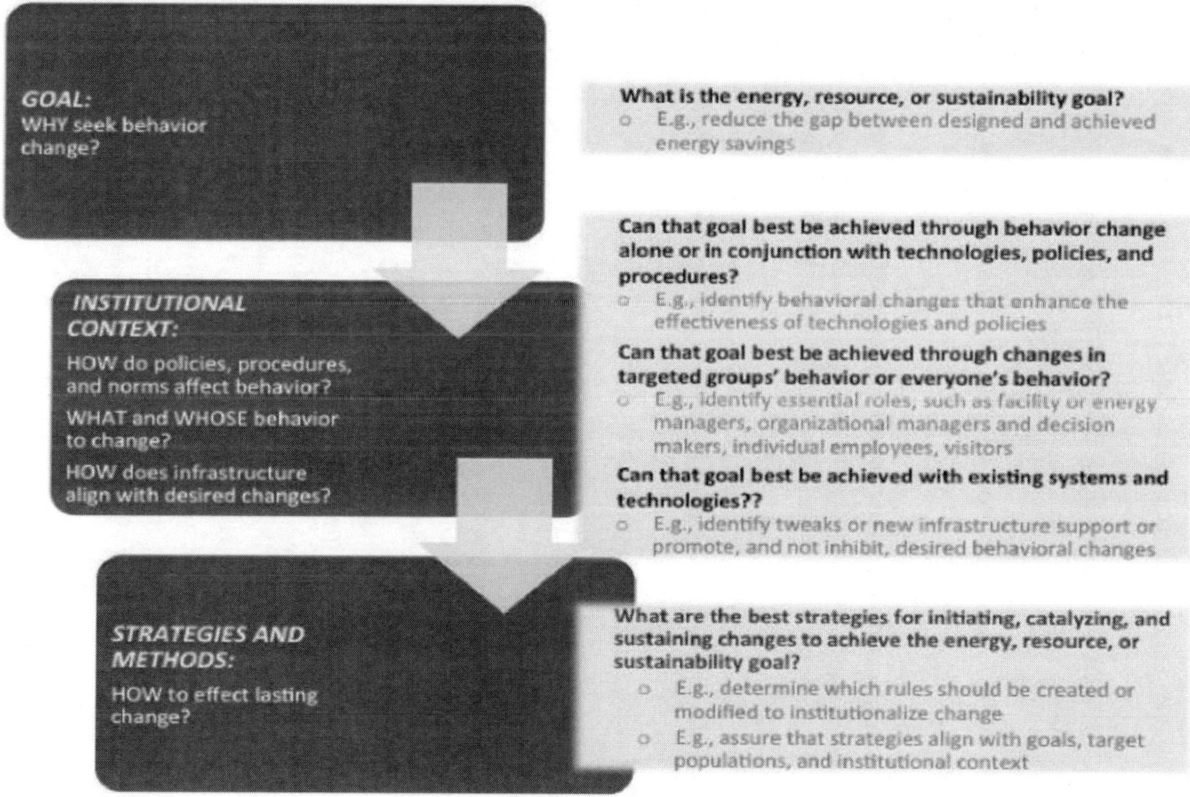

Figure 4.1. Process Diagram for Identifying Institutional Change Elements (Rules, Roles, and Tools) that will Drive the Selection of Strategies for Achieving Change Goals.

The construct of "rules, roles, and tools" can help to structure an *Action Plan for Behavior Change* to achieve sustainability and building performance goals. Briefly, the terms can be described through the questions they address:

- **Rules**: What are the policies, procedures, and norms that support the present behaviors and the principles that apply to changing those rules?
- **Roles**: What/whose behavior matters with regard to that problem? What needs to change?
- **Tools**: What technologies, processes, and systems are in place or needed to support desired behavior changes and outcomes?

Rules, roles, and tools are inextricably linked and are key components of institutional context. It is crucial to understand both the formal and informal rules that shape the way organizations and individuals within them operate, whether those rules take the form of formal executive orders establishing Federal agencies' sustainability goals; procedures for maintaining a building's mechanical equipment; or more informal, tacitly accepted ways of doing business. New or modified rules may constitute effective methods for ensuring that sustainability-oriented behaviors endure in organizations, outlasting staff turnover and dimmed recollections of why formal or informal rules exist. It also is essential to identify categories of people within organizations who have the authority to effect change or whose behavior (role) is critical to the success of a technology or policy. At the same time, revising job descriptions is among the straightforward strategies available to initiate lasting change in organizations. Tools, then, are the technologies, processes, and systems that constitute an organization's infrastructure. That infrastructure may serve to enhance or inhibit behaviors that align with sustainability goals. As an example, procurement systems may be designed in ways that make it easy—versus difficult—to select "green" products (and hard not to select such products). Similarly, easily accessible duplex printers are essential for complying with directives to reduce wastes by printing on both sides of paper.

Once an organization's rules, roles, and tools are assessed, it is possible to select suitable strategies and tactics for achieving and sustaining change in organizations. It is vitally important to select strategies and tactics that align both with energy or sustainability goals and with the institutional context of an organization (rules, roles, and tools).

In taking this institutional context approach, rules, roles, and tools distinguish among sustainability and performance goals, building-related

O&M challenges, and key players, parties, and strategies. Mapping the building ecosystem in this way reduces the complexity of this seemingly intractable concept called "behavior" or "behavior change" and specifies ways in which evidence-based, proven approaches can make a difference.

Importantly, significant and persistent change requires multiple interventions at multiple levels over time — that is, changes relevant to rules, roles, and tools — as the extent to which change needs to be supported by other changes becomes known and feasible. Heath and Heath (2010) point out that "moving change forward is best accomplished by searching for both successes and problems, then using the successes to overcome the problems." This insight provides a simple construct to help identify a path forward – using in situ successes and failures in the dynamic building environment.

Setting a clear goal is the first step in the process of achieving higher-order sustainability goals. Once the goal is established, it is important to assess the institutional context within which that goal is to be achieved. In the section below, we delineate the rules, roles, and tools components of institutional context, providing specific examples.

4.1. Rules

> Individual, group, and institutional rules, both formal and informal, directly influence occupant behaviors and must be identified and addressed.

Action: Identify the formal and informal rules that support current behaviors targeted for change and that influence the organization's ability to achieve the specified sustainability goal.

Formal and informal rules shape behavior in the workplace, establishing what constitutes preferred, acceptable, and unacceptable actions. Formal rules can be interpreted in multiple ways. In practice, it is the informal rules that often determine how formal rules are interpreted and, should multiple rules apply, which rules actually should be followed.

Rules also can be used to prompt and reinforce sustainability-oriented behavior changes. For instance, short-term change may be achieved through strategies such as campaigns, competitions, and training. But for change to "stick" and sustainability goals to be achieved over the long term, new behaviors must be supported by new rules. Such rules include policies, procedures, and informal practices. Initial efforts to identify needs for new rules or modifications of existing rules should be goal-oriented, linking to resource-use or GHG reductions; and context-specific, taking into account—and, ideally, having neutral or positive impacts on—mission and job responsibilities.

Example: In a situation where the goal is to reduce the gap between designed and achieved energy savings, one might ask about the actual maintenance schedule, known operating problems, and observed occupant behaviors (e.g., leaving exterior doors open, blocking vents, using individual coffee pots or space heaters). After diagnosing the problems, a planner or analyst can identify appropriate adjustments. Leaders can alter policies or practices, and lead by example, e.g., by visibly engaging in conservation activities or actively communicating new procurement procedures. Can defaults be reset on computers, printers, and other office equipment in ways that occupants see as contributing to, rather than impeding, their ability to do their jobs? Can building policies be changed? Can communal devices such as coffee pots or refrigerators be made more appealing to use as substitutes for multiple personal devices? To analyze and determine effective change options, a review of formal rules needs to be supplemented by dialogue with people who represent critical-path roles and who may be most affected by altered policies, technologies, or practices. This dialogue can identify non-obvious rationales for behaviors and habits, and thus may lead to potentially effective strategies for creating and promoting new habits. The effectiveness of these strategies then can be tested.

Evidence: Smith (2005) suggests that managers should alter the environment, context, or infrastructure to make change easier than continuing old behaviors. Managers should problem-solve with staff and remove barriers and change-conflicting policies and procedures (Goins and Moezzi 2011, 2013). Prindle (2010), in his studies of successful energy-efficiency efforts in firms, reiterates that firms identify staff involvement as an essential element to effecting change.

4.2. Roles

> It is important to determine the roles that are essential in achieving a goal and who, in which roles, can effect change that will solve the problem.

Action: Analyze who is essential and who can make a difference in a specific problem area.

Example: If the problem goal is to shrink a documented gap between as-designed or benchmarked energy use and higher actual energy use, it will require several categories of people to help close that gap (Cole 2011). Examples include the following:

- Building managers and O&M staff, who have perhaps the largest role in saving energy: they know the building systems and can uncover technological issues and observe occupant behavior that works against reduced resource use.
- Occupants, who, in the course of performing their jobs or using buildings, may increase resource use by resetting temperature controls, blocking vents, adding personal small appliances or lights, and so on. Because of the focus on occupants in many sustainability programs, there are many potential behaviors that could be changed. However, it is important to make sure that (1) occupants can control resource use in the ways specified ("tools"); (2) their legitimate needs for comfort, indoor air quality, and lighting are achieved; (3) their job-related needs are met; and (4) new or modified behaviors become embedded as "normal" so that they persist and remain effective as situations change (e.g., new responsibilities, technologies, or staff members).
- Leaders, from top management to staff champions, who visibly advocate and set examples of energy-saving behavior. Leaders and champions may overlap with other roles.
- Managers and technical representatives, who approve purchases and thus can specify ENERGY STAR®, Federal Energy Management Program-approved, or other resource-efficient products. These individuals may have the authority to limit the number and types of personal equipment, as well as the (sometimes conflicting) responsibility to provide workspaces that enhance productivity and comfort.
- Contracting officers and procurement personnel (i.e., major "back office" players on the actual products and services purchased), who can also limit their purchases to those that conform to energy, water, materials, and indoor air quality standards.
- Information technology specialists, who can set defaults for "sleep" modes, double-sided printing, etc., and encourage overnight shutdown

of equipment (e.g., by applying patches at start-up or shut down), while assuring these practices align with other organizational responsibilities (e.g., cybersecurity).

- Administrators who set building policies such as hours of operation, lighting protocols, and office assignments (taking into account such factors as occupancy rates).

It is extremely important to differentiate the roles that are involved in the behaviors targeted for change, and to engage the pivotal roles first — those with the leadership, authority, resources, and influence to initiate and maintain changes that will lead to reaching sustainability and building performance goals, as well as those whose "buy-in" is necessary. Once pivotal roles are aligned and individuals are motivated, a comprehensive change program can be developed (Gardini et al. 2011).

Evidence: The literature on organizational change (e.g., Aiken and Keller 2009; Harvard Business Review 2011, 2001; Kotter 1996; Argyris and Schon 1996; Lewin 1951) emphasizes that multiple roles and levels need to be involved to both effect change and to institutionalize it. Smith (2005) finds that asking people to do something they cannot do makes the situation worse.

4.3. Tools

> It is essential to assure that the organization's infrastructure is in alignment with the behavioral changes necessary for achieving sustainability objectives.

Action: Assess the degree to which existing technologies, systems, and processes support or inhibit the ability to achieve sustainability and energy-use goals.

Example: Continuing the example above, knowledge assembled about rules and roles — especially when gleaned from dialogue with relevant staff — can help to identify the extent to which the organization's infrastructure needs to be tailored to effect behavior changes. Only when tools are appropriate and properly aligned with goals, rules, and roles can they effect and sustain change. Examples include revising maintenance practices or other standard operating procedures, recommissioning buildings, embedding specific energy-efficiency requirements within procurement procedures, and optimizing equipment settings and operation. Other tools reflect changes in how business is conducted — a new rule to discontinue the use of space heaters could be combined with infrastructure changes such as a new process through which building managers improve comfort levels or the installation of new occupant-responsive technologies. Coupling these tools to promote both short- and long-term change can help ensure that sustainability successes persist once they are achieved. If formalized rules (e.g., prohibitions on individual space heaters) and standard operating procedures (e.g., timely responses to complaints about discomfort) are enacted, rules can become tools.

4.4. Strategies to Prompt and Sustain Change

It is essential to identify which strategies are best-suited for achieving sustainability goals within the rules, roles, and tools of an organization that will best facilitate occupant behavioral changes to achieve sustainability objectives. The set of strategies that constitute the behavior and behavior change "portfolio of options" are not interchangeable; sustainability managers must carefully match strategies to situations, problems to be solved, and feasibility for users. Strategies must be selected for specific contexts to achieve specific goals over specific periods of time (e.g., how long might commitments be relevant?). A hammer and drill serve different purposes, and so do formal standard operating procedures, competitions, and real-time feedback. Likewise, just as both hammers and drills are essential in building construction, multiple behavior and behavior change methods may be necessary to achieve an enduring organizational culture of sustainability.

Some modifications to existing rules, roles, and tools are particularly suited for institutionalizing change. Examples include rewriting position descriptions and performance standards, requiring the reporting of progress toward goals, instituting required training of new rules and strategies, changing the organizational structure, and adding budget lines to highlight costs/savings or to support initiatives. These types of strategies promote a shift from implicit or short-term expectations about sustainability to explicit, long-term ways of doing business.

Evidence: Evidence-based approaches for behavioral change exist and are provided in Sections 5 and 6, with the caveat that many have not been tested in rigorous ways in Federal or commercial office settings or for sustainability goals, or for organizations (as opposed to individuals).

5.0. Strategic and Specific Methods to Achieving Behavioral and Institutional Change

This section identifies specific methods that have been used as behavioral interventions. It notes some of the virtues and limitations of different methods, including contexts in which methods may be more or less appropriate and effective. It draws primarily from the body of psychosocial literature, which seldom addresses institutional change or sustainability-related behaviors. For these reasons, this section begins with a short discussion of that literature as a whole and, throughout, identifies uncertainties and gaps in applying the findings to Federal building sustainability goals.

5.1. About Peer-Reviewed Behavioral Change Literature

Research on behavioral change distinguishes between interventions aimed at individuals or groups (psychosocial interventions) and those aimed at institutional policies and practices (structural interventions). While the previous section focused primarily on structural changes, this section focuses on psychosocial interventions to support and reinforce changes in policies and practices.

The different levels of focus (individual behaviors versus structural changes) derive, in part, from the academic discipline of the researchers. Cognitive and social psychologists typically focus on the mechanisms influencing action and thus tend to have tightly constricted research paradigms and relatively short time frames for conducting research. In contrast, sociologists and anthropologists generally take a broader view of behavior that links culture, roles, and other contextual factors. This paper explicitly incorporates both approaches as necessary to achieve persistent change across an organization.[1]

The majority of research discussed in this section draws on psychological frameworks and involves mechanisms that act on individuals or small groups. Topical areas divide this section. For areas with a rich and extensive body of research, it is possible to identify best practices. For other emerging areas, such as behavioral economics (e.g., Ariely 2009; Thaler and Sunstein 2009),

[1] For further detail and resources, see the U.S. Department of Energy Federal Energy Management Program's Institutional Change website: http://energy.gov/eere/femp /institutional-change-sustainability

there is not yet a sufficient body of research on building-related behaviors to be able to identify consensus best practices. In these cases, we briefly describe what is known and its potential application to building-related behavioral change.

5.2. Providing Information and Outreach

> *Information* and *outreach* are systematic attempts to provide important knowledge beyond standard norms to particular segments of a community, with the intent of initiating change.

Information and outreach efforts are among the most common approaches to encouraging behavior change and have been widely used in public health campaigns (Campbell et al. 2000) and in initiatives oriented toward home energy use (Abrahamse et al. 2005). As in the public health arena, information and outreach often are used in the hopes of prompting individuals (as distinguished from organizations) to change their behavior. Information (education) campaigns may provide general material about environmental problems (e.g., impact of energy and water use on the environment) or may contain specific recommendations on how to save energy and water. Techniques for information and outreach include workshops, mass media campaigns, videos, and direct mail. Sometimes techniques are highly tailored, like energy audits. Providing information in various formats is usually necessary, but information alone is not sufficient to effect change (McKenzie-Mohr and Smith 1999).

Campbell et al. (2000) conducted an extensive review of the effectiveness of community-based public health interventions that used a wide range of strategies including educational sessions, curriculum development, networking and coalition development, counseling, mass campaigns, and home assessments. The review found that many of the projects were effective in increasing short-term public awareness of environmental health risks, but that few were associated with actual changes in behavior. The most successful programs used multiple awareness events and multiple settings (see Section 6.2, *Appealing to Multiple Motivations*).

Although information and outreach are commonly used approaches, they may not be the most effective methods for instigating or sustaining behavior change. Campbell et al., for instance, found that even if information reaches its targeted audience and affects attitudes, it does not necessarily lead to behavior change. That is, it is possible to "change minds" without changing behavior. This finding may be especially true when the specific behaviors targeted for change are well-established habits[2], as noted by Verplanken and Wood (2006, p. 91), who state, "consumers' everyday lifestyle habits limit the effectiveness of interventions that do not address the performance contexts and social structural factors that maintain habits."

However, because information and outreach can successfully increase knowledge and possibly change attitudes, they may be important first steps in many behavior change programs. The Stages of Change model (Prochaska et al. 1997) uses information and outreach to increase awareness of a problem and to "plant the seeds" for change. Thus, change managers need to know what factors increase the effectiveness of outreach messaging in raising awareness.

Success factors and best practices for information and outreach messaging (remembering that success here means "message received") include reviews of successful outreach campaigns. Following are success factors and best practices for information and outreach messaging as derived from the literature (e.g., Abrahamse et al. 2005; Campbell et al. 2000) which shows messages are more likely to be seen and remembered when they:

- Provide vivid, concrete, and personalized information.
- Use credible, trustworthy sources to convey messages.
- Use clear, concise, and concrete language to increase understanding and memorability.
- Use multiple methods and settings, and deliver content over multiple sessions rather than during a one-time effort.
- Tailor the information to the specific context.
- Circulate information through social networks.

---

[2] Actions to reduce cigarette smoking are a well-known example from public health. For nearly a half-century since the publication of a pivotal U.S. Surgeon General report in 1964, there have been continuing information campaigns together with the enactment of rules pertaining to advertising and restrictions imposed in public- and private-sector establishments. These experiences attest to the (a) failure of information, knowledge, or self-interest automatically to lead to behavior change; (b) need for multiple strategies; and (c) possibility that achieving an enduring change may require long-term, continuing efforts.

5.3. Giving Feedback

> *Feedback* techniques provide information on how well an individual or group is performing relative to a specified goal, how current behavior outcomes compare to past outcomes, or how personal behavior outcomes compare to those of other individuals or groups.

Feedback in one form or another has been a common element of behavior change initiatives for decades, especially for residential energy use. Recently, the means of delivering feedback has expanded. Now there is the ability to see links between specific behaviors and progress toward achieving goals in real time, highlighting the integration of new technologies with social systems to provide highly accessible and immediate feedback, e.g., on a dashboard display in front of the viewer. The key challenge for feedback approaches is to use data not only to inform, but also to motivate change. Feedback by itself is only information; how people respond to the information is the focus of behavioral change efforts.

When applied to energy consumption, primarily in residential settings, feedback techniques have resulted in energy reductions of 5 to 12 percent depending upon the specific context and methods used (Froelich 2009; Ehrhardt-Martinez 2010). Furthermore, studies of continuous versus monthly feedback about energy usage show that continuous feedback is more effective in achieving reductions in energy use. Households receiving continuous feedback reduced gas use by 12.3 percent compared to 5 percent for households receiving monthly feedback (van Houwelingen and van Raaij 1989). Other studies find similar results (Foster and Mazur-Stommen 2012). In response to OPower's Home Energy Reports sent to Minnesota electricity customers, there were verified reductions of 2.10 to 2.27 percent in the first year of the program (Power System Engineering 2010). In another study (Winkler and Winett 1982) based on 19 sets of data from experimental studies, households were informed frequently (usually daily) about how much energy they were using. Study participants who modified their habits reduced their energy consumption, some by as much as 20 percent. Darby (2006) reviewed 38 studies of energy efficiency and found that feedback reduced energy usage by 10 to 14 percent on average.

Frequent or real-time feedback can now be provided through smartphone applications (apps), equipment sensors, and computer- or lobby-based dashboards, and persuasive technologies can be applied to social groups

(Consolvo et al. 2009). The technologies now available may be especially useful for energy behaviors insofar as they have the potential to solve the "invisibility" problem (Froelich 2009). Because energy is invisible and untouchable, it is especially difficult to motivate behavior change and to identify links between behavior and energy use. Several technologies focus on "seeing" energy. Dashboards, for instance, can be designed to display graphs of real-time resource use of appliances, trends over time, differences between groups, and other relevant data analyses.

Other technologies show energy in a more abstract manner and are intended to send rapid and easily recognizable messages. The idea is to make energy use available when and where the user needs it without interrupting ongoing activities. For example, a lighted energy orb in one application glows green when energy use is within the desired limits, but glows red when energy use is high (Consolvo et al. 2009). This method does not indicate how much energy is being used, but whether it is within or outside set targets. Feedback on progress toward goals is a common element in many applications. These applications range from simple graphics or statements about progress to emotionally appealing rewards (Consolvo et al. 2009). For instance, a phone app related to exercise uses a garden analogy. As exercise increases (as measured by a pedometer), the display shows flowers growing. As progress toward the goal increases, more flowers appear in the garden. Another exercise-based program described by Consolvo and colleagues enables users to exchange information on their progress with others and to provide encouraging messages. Another activity-based application uses a desktop "statue" linked to physical activity levels of workstation occupants. When activity levels are within the set goals, the statue is upright. When activity levels are too low, the statue slumps. Although these approaches do not currently focus on resource consumption, the potential exists to provide feedback on energy or water use.

Most of the literature focuses on individual behavior and on behavior changes from which individuals derive tangible, personal benefits. However, some literature indicates that feedback also can be effective in social groups. Siero et al. (1996) combined feedback with social norms (see Section 5.8, *Tapping into Social Norms* below) in an experiment designed to reduce energy consumption in an office building. One test group received feedback only about its energy consumption relative to a set goal, while two other groups received comparative feedback — how well each group was doing relative to the other. The experiment was run over a 20-week period from November through March. Research methods included assessment of attitude change as well as actual energy reduction. Results showed that the comparative feedback groups were more successful in saving energy than the group that received feedback only about its own performance. Furthermore, those in the comparative groups also changed related behaviors that were not targeted. Interestingly, the experiment showed no changes in attitudes among all three groups over the study period. It also is unclear how long the observed behavioral changes would persist in the same groups, or as group membership evolves over time.

Following are success factors and methods with evidence of success for feedback as derived from the literature:

- Provide more frequent (daily, real time) versus less frequent (weekly, monthly) feedback.
- Make data available in an unobtrusive manner when and where needed.
- Combine feedback with specific actions to reduce energy consumption (or other sustainability outcomes).
- Provide feedback on how well people are doing, such as progress toward goals and goal achievement.
- Reinforce or reward successful resource-saving behaviors, which can include verbal praise or visual methods that translate impacts of behavior into environmental benefits (more trees, growing gardens).
- Combine feedback with prompts, reminders, or other triggers to help promote and maintain desired behaviors.

5.4. Persuading

> *Persuasion* is the influence of beliefs, attitudes, intentions, motivations, or behaviors. Persuasion techniques go a step beyond information and outreach in that they explicitly aim to convince people to take certain actions. With persuasion, there is no assumption that "the data speak for themselves" (i.e., that information will lead to action).

Persuasion methods seek to convince people to change their behavior more explicitly than either information or feedback. One well-recognized source on persuasion and behavioral change is Robert Cialdini's six "weapons of influence" (2007):

- **Reciprocity:** the strong cross-cultural norm of returning favors. Can reciprocity be used to encourage resource conservation? The current focus is on competition between groups rather than reciprocity to achieve environmental goals (see the titles of numerous papers from the Behavior, Energy and Climate Change conferences[3]; also Metzger et al. 2011).
- **Commitment and consistency:** the tendency to honor an idea or goal that has been committed to orally or in writing, even if the original incentive is removed. Commitment strategies have been used since the 1970s oil embargo to reduce energy consumption.
- **Social proof:** the tendency for people to do things others are doing. This driver relates to social norms and conformity to group behavior. Social norms are increasingly used to generate behavioral change in organizations and communities.
- **Authority:** the tendency to obey authority figures, which is relevant to organizational change and the role of leadership in motivating and sustaining change.
- **Liking:** the tendency for individuals to be easily convinced by people they like, which is relevant to the selection of "messengers" in a behavioral change program.
- **Scarcity:** in marketing research, scarcity tends to generate demand ("limited time only"). Of relevance here is how scarcity of environmental resources can be used to *diminish* demand and lead to more efficient use (such as reduced electrical use during high-demand periods).

5.5. Offering Rewards and Incentives

> *Rewards and incentives* serve to induce or motivate behavioral change.

Rewards and incentives have been frequently used to motivate and sustain behavioral change, from tokens for bus ridership to coupons for purchasing green products. Several recent projects illustrate the value of incentives. Project Porchlight[4] (run by One Change, a nonprofit organization in Ottawa,

---

[3] See http://peec.stanford.edu/events/2011/becc/ and http://peec.stanford.edu/events/2010 /becc/ for the most recent conference papers.

[4] See http://www.projectporchlight.com/.

Ontario, Canada) uses social networks to distribute compact fluorescent light bulbs (CFLs) and encourage households to change their behaviors. In another activity, McKenzie-Mohr and Smith (1999) sent a pamphlet on energy conservation to two groups of homeowners. One group in their study received a shower flow restrictor along with the pamphlet; the other group received just the pamphlet. Homeowners who received the shower flow restrictor were more likely to engage in the other conservation actions mentioned in the pamphlet (e.g., reducing the temperature on their hot water heaters, installing setback thermostats, and cleaning their furnaces). The CFLs and shower flow restrictor programs are also examples of Cialdini's "reciprocity" principle in which people who receive a favor or gift feel socially bound to reciprocate in some way.

A careful selection of items that serve as rewards or incentives is important; some may have the unintentional effect of demotivating the people they are designed to motivate. Monetary awards, for instance, are not motivating in the ways that people often assume. Pink (2010) and Ariely (2009), for example, describe situations in which people will be motivated to work hard without monetary compensation because of friendship ties, the ability to be creative, or their belief in a larger cause.

However, offering the same people money places the relationship on a business level — in which case, compensation can seem inadequate and demotivating. Pink suggests that money may act as a disincentive for creative, versus mundane, tasks. Ariely points to the difference between social (familial, friendship, romantic, in-group) norms and market norms; when they collide, "trouble sets in" (Ariely 2009, p. 71). Ariely also reports on an experiment in which people who performed the same task were more productive when unpaid than when paid, just as lawyers are more willing to take on pro bono work than poorly paid (albeit paid) work.

5.6. Changing Defaults, Nudging Choice, and Enabling Adaptive Comfort

> *Changing defaults* — How building space is configured and how options are presented can make huge differences in the choices people make.

Sustainability can be promoted through good *choice architecture* — organizing a context in which people make decisions and shaping the range of behavioral options available to them (Thaler and Sunstein 2009). The concept of choice architecture was developed with reference to decisions that individuals may make in their own self-interest, such as healthier dietary choices, greater retirement savings, and the like. When applied to the realm of energy or sustainability, there is an attendant shift from individual self-interest to organizational or national interest. The idea is to encourage people to make choices that, in the case of this paper, achieve Federal sustainability objectives. In this context, "default" options have been the predominant method used to create or alter choice architecture. Examples of default options are preset thermostat settings and the order in which "green" supplies are presented in procurement software.

Identifying "appropriate" or effective default settings may require assessment. For instance, the default settings of the physical environment can either promote or hinder energy-saving behavior. They need to make sense in the physical and workplace contexts in which they are used; a single default setting may be inappropriate for an entire building or organization. For example, the same temperature setting may result in different levels of comfort on north- versus south-facing sides of buildings or in workspaces where occupants are sedentary versus active. Building managers also can widen the default temperature range and give occupants leeway to adapt. Advantages accrue in lower energy demand, higher staff satisfaction, and ease of operation. (Buildings/people are complicated enough that attempts to narrowly control temperature may often work poorly.)

Similarly, automatic lighting controls that go on when people enter daylit spaces can waste energy. In these settings, a more effective default practice may be to provide the option to turn on the lights but have automatic controls turn them off when people leave the room (although people who turn lights on may be more likely to remember to turn them off).

Another example of choice architecture through default options involves purchasing decisions. If energy-efficient products are the default selection, rules could be enacted that would require a purchaser to justify making a non-sustainable choice. This approach is markedly different from one in which sustainable items are an option merely identified by word or icon in a product catalog. Choice architecture can also play a role in new equipment decisions; in 2007, Walmart placed CFLs in the most visible, central, and accessible positions on its shelves and thus encouraged sales of energy-efficient products (DOE 2009).

Moezzi (2009) discusses an interactive process of "adaptive comfort," which provides more choices for people to adjust their behaviors or the environment as they desire. Adaptive comfort could include mobility supports (laptops, wireless connectivity, mobile phones, etc.) that enable people to move to different spaces as work or comfort needs change. It also provides more choices within individual workstations, such as control over ceiling lights, task lighting, and ventilation.

Researchers with the Center for the Built Environment at the University of California, Berkeley, report on developing and testing technologies to support adaptive comfort needs (Zhang et al. 2009). These technologies include small energy-efficient fans and hand- and foot-warmers designed to condition the body envelope, while enabling ambient temperature conditions to fluctuate depending upon external weather conditions. In a laboratory test chamber, researchers found that they could maintain comfort levels across a wide range of temperature conditions, from 64 to 86 degrees Fahrenheit. They also projected potential energy savings of up to 40 percent from reduced ambient heating and cooling loads.

Beyond the few illustrations presented here, it would be useful to rethink how options are created and implemented in buildings through physical design, building operation and maintenance, and workplace procedures and practices — individually and interactively. Examples range from the location of doors, stairs, and elevators to the use of light at night for security and cleaning purposes. For instance, the design and location of stairways strongly influences whether people will choose to use the elevator or the stairs (Nicoll 2006). For energy (and health) purposes, encouraging building visitors and occupants to use the stairs may be highly desirable.

5.7. Gaining Commitment

Commitments are oral or written pledges to change behavior.

Commitments have long been used as methods for instigating behavioral change. Such commitments are often linked to specific goals, such as decreasing energy use by a certain percent (Abrahamse et al. 2005). A study by Katzev and Johnson (1984) analyzed the impact of commitment on home energy use. Households in the study who signed a commitment to reduce their home energy by 10 percent saved more energy than the control group. A similar study assessed the effectiveness of public versus a private commitment (Pallak and Cummings 1976). The households in the public commitment group saved more energy than those who made the private commitment or who were in a non-commitment control group, and the behavior was sustained over six months.

In another experiment, home assessors in California were trained to make use of commitment strategies and persuaded three to four times as many people to elect to retrofit their homes compared with other home assessors. The commitment-trained assessors sought verbal commitments from householders: for example, an assessor might ask a homeowner when he or she anticipates having a weather-stripping project completed and suggest a call around that time to gauge progress and/or offer help (Hirst 1984, cited in McKenzie-Mohr and Smith 1999).

Public commitments are especially powerful, because they activate social norms of commitment to others and may also activate reciprocity norms and values. However, the degree to which they engender lasting change remains unknown.

5.8. Tapping into Social Norms

> *Social norms* are the explicit or implicit rules specifying what behaviors are acceptable within a society or group.

Social norms spur and guide behavior. As a social species, humans are programmed to want to be accepted by peers and to conform to their behavioral practices. This concept has been increasingly used in the environmental field in interventions designed to reduce energy consumption and to increase recycling. In some cases, these interventions use information or feedback to reveal otherwise hidden norms to promote desired behaviors. In other cases, appeals to social norms have been used as ways to guide people to make desired choices in situations where they otherwise may not know what to do. However, research has shown that the social norms approach is not always successful and that reactions vary when people learn that their behavior is different from the norm.

For instance, in research on residential energy use, households received bills showing their energy consumption compared to others in their neighborhood (Schultz et al. 2007). Households that were above the norm tended to reduce their consumption; however, households below the norm tended to increase their use. This latter behavioral response clearly increased conformance with the average consumption, but it was also undesirable and constituted a "boomerang effect." Researchers then added another condition to see if they could reverse the boomerang effect and maintain the lower level of consumption among the households who used less energy than the norm. To do so, they added a visual communication element to bills intended to provoke an emotion-based behavioral response. Smiley-face emoticons were placed on the bills of households that were below the norm, while households that were above the norm received bills with frowning face emoticons plus hints on how to save energy. The faces were intended to provide a message of social approval or disapproval — what Cialdini (2007) refers to as an "injunctive" norm. The research showed that households that received the frowning face significantly reduced their energy use compared to the baseline; households that received the smiley face maintained their higher levels of conservation and did not regress toward the norm.

In another example, Griskevicius et al. (2008) studied the use of towels in hotel rooms. These authors point out that, when a situation changes and people do not know what to do, they look to others for evidence of how to act and will be attentive to examples of new behaviors, especially if those behaviors are specific to the changed situation. They found that guests who received a sign saying that the "majority of hotel guests in *this room* reuse their towels" (italics added) were more likely to reuse towels than guests who received the traditional appeal to save the environment by reusing towels. The message linking towel use to "this room" was also more successful at changing behavior than a message saying the "majority of hotel guests" reuse their towels. The closer the reference group is to the hotel guest (e.g., staying in the same room), the more powerful the influence on behavior. This finding supports Cialdini's (2007) "weapon" of social proof — the tendency to mimic the behaviors of others, especially in novel or changing contexts.

5.9. Social Marketing in Communities

> *Community-based social marketing* is a programmatic approach focused on barriers to and benefits of changing behavior at the community or group level.

Another promising method combines psychological and sociological approaches with community-based social marketing (CBSM) and is especially relevant to behavioral change in buildings. Organizations and occupants of buildings are both communities of people and can thus benefit from methods that use a community focus. CBSM uses multiple methods to move communities toward sustainable resource use (McKenzie-Mohr 2009; McKenzie-Mohr et al. 2012). The methods commonly employed in CBSM are commitment, appeal to social norms, communications that frame and deliver the message to targeted audiences, incentives that motivate new behavior, and prompts to serve as reminders. The specific approach developed by McKenzie-Mohr (2009) identifies the barriers to and benefits of changing existing behaviors to more desirable future behaviors. This emphasis on barriers and benefits, identified through research reviews as well as public meetings, enables a program team to use different strategies to overcome barriers to change while simultaneously supporting desired new directions. The focus is on voluntary individual behavioral change, not organizational behavior change.

Gershon (2009) also uses multiple methods, including peer-to-peer support groups called EcoTeams; commitments from local governments, faith groups, and environmental organizations; and promotion of co-benefits, including more livable neighborhoods and economic development. In the long term, these strategies build community leadership and political support to sustain the changes.

A study of the effectiveness of two CBSM programs found generally high effectiveness in changing behavior, but also variability across behaviors (Pickens 2002). The "EcoTeam Program" in Deschutes County, Oregon, focused on reducing household resource consumption using the CBSM model. An analysis of the program showed that households taking part in the program significantly reduced their water and energy use and garbage output, but not their transportation behavior. In contrast, a program in Boulder, Colorado ("Go Boulder") that focused on reducing vehicular miles traveled found a six percent shift toward use of public transportation. These findings suggest that some behaviors, like household resource use, may be closely linked, but that others, like transportation, are sufficiently different that they need a specialized focus.

CBSM approaches may be especially valuable in shifting building-related energy use behaviors. Buildings and organizations within them are communities in much the same way that neighborhoods are communities. The CBSM approach provides multiple means of engaging the building occupants and incorporates their input into the selection of methods and messaging for specific audiences (i.e., roles). The approach can also be used at the organizational level to identify barriers to and benefits of changes in rules and roles.

5.10. Responding to an Urgent Situation

> *Urgency* is a quality or condition suggesting immediate or pressing importance.

When people experience a crisis they deem real, they act. Juneau, Alaska, when faced with the destruction (by avalanche) of a main power line in 2008, cut its electricity consumption by 25 to 30 percent in a few weeks (see Pasquier 2011 for this and other examples of "saving electricity in a hurry"). Severe energy austerity was self-imposed by the U.S. Army bases in Japan impacted by the earthquake and tsunami in 2011.[5] However, examples such as these are rare, highly situational, driven by dramatic mergencies, and not long-lived. A sense of urgency need not derive solely from catastrophes or disasters, however. Important deadlines, potential penalties, public scrutiny, economic competitiveness, and other circumstances can drive a sense of urgency. There is limited evidence that residual savings may occur if the "urgency" is reframed from one of environmental danger to lost economic opportunity (Haanaes et al. 2012). The urgency may then spur innovation, productivity, and economic thriving. A sense of urgency may be more valuable in focusing and directing change efforts than in sustaining them over time. For instance, one of the most widely used approaches to organizational change, developed by John Kotter (1996), cites "establishing a sense of urgency" as the first of eight steps for transformational change. Among the actions needed to establish urgency is "convincing 75 percent of your managers that the status quo is more dangerous than the unknown." Two barriers to establishing urgency are underestimating the difficulty of moving people from their comfort zone and risk avoidance (or as Kotter says, "becoming paralyzed by risks").

6.0. Using Cross-Cutting Elements in Selecting Methods for Achieving and Sustaining Behavioral Change

Social science literature sometimes addresses aspects of behavior change that do not fit neatly under the headings of rules, roles, and tools. This section addresses two such topics (habitual behavior and motivation) because of their

---

[5] Personal communication from Richard G. Kidd, Deputy Assistant Secretary of the Army (Energy and Sustainability), U S. Department of Defense, 9/26/2011.

underlying importance in understanding, designing, and implementing behavior change interventions that enhance overall building performance. It also discusses the persistence of behavioral change, understanding the difficulties in evaluating such changes and their energy- or sustainability-related impacts over long periods of time.

6.1. Establishing Habits that Foster Sustainability

> *Habits* are actions or activities that are carried out automatically, with little conscious effort, and are often cued by factors in the environment.

The vast majority of everyday behaviors are habitual, automatically triggered by contextual cues (locations, previous actions) that fall below the level of conscious awareness (Wood and Neal 2009). Many of the behaviors of interest to this paper — especially energy, water, and technology use — are likely to be strongly habitual. As noted by Wood and Neal, "In daily life, the tendency to act on habits is compounded by everyday demands including time pressures, distraction, and self-control depletion." Although difficult, undesirable habits can be broken; they will be replaced by new habits as soon as the context provides consistent, regular cues to support new actions. A compounding difficulty is that people have a limited pool of willpower to draw on, so enabling conditions are needed. The challenge here is to break unsustainable habits and replace them with pro-environmental alternatives, and then to support the new habits with suitable cues (Duhigg 2012).

Importantly, individuals' habits change as contexts change. Because habits are cued by context (physical environment, the social-institutional environment, existing policies and procedures), a change in the context provides an opportunity to intervene in creating new habits (Goins and Moezzi 2011, Verplanken and Wood 2006, Wood and Neal 2009). Moving to a new or renovated building, for instance, could provide the opportunity to intervene and shift habits by using many of the methods described in this study, from goal setting and commitment to rewards and motivation. Wood and Neal identify several conditions that are most conducive to forming new habits:

- A stable context in which the behaviors are performed frequently
- Rewarding behavior — in early habit formation, greater rewards lead to repetition of the behavior (necessary to create a habit)

- Behavioral reminders in strategic locations, such as recycling signs at trash cans
- Goal setting to increase the intention to change, which is a precursor of actual change

Habits also can change when individuals exert self-control. In a study of people's attempts to change their habits (Wood and Neal 2009), the most successful strategy was "vigilant monitoring" described as "heightened attention to and inhibitory focus on the unwanted response (e.g., thinking 'don't do it')." However, effortful self-control also involves significant mental and emotional effort to identify the context triggers, the full pattern of the habitual behavior, and the specific reward that reinforces the behavior, as well as the effort to maintain change.

For sustainability purposes, changes in the building or organizational context may be more promising ways to create new environmental behaviors than relying on occupants to exert individual self-control.

6.2. Appealing to Multiple Motivations

> *Motivation* refers to a process that elicits, controls, and sustains certain behaviors.

Most intentionally promoted behavior change requires several stimuli, or "nudges" (Thaler and Sunstein 2009), from external sources. Psychologists view motivation as one of the key points of leverage for behavioral change. Fortunately, motivation comes in many forms.

Weber and Lindemann (2007) point out that decisions are made (1) "by the head" — calculation-based decisions, (2) "by the heart" — emotion-based decisions, and (3) "by the book" — rule-based decisions. Heffner et al. (2006) examined why people bought hybrid electric vehicles early in their introduction to the mass marketplace, since the payback period was so long. They concluded that individuals received symbolic meaning and value from owning such vehicles (a "by the heart" decision). In a classic study of organizational culture, Schein (1999) affirms the value of demonstrating that desired changes reinforce existing assumptions about the way to act. In an analysis of multiple motivations in green buildings, Miller and Pogue (2009) found that tenants in green buildings tend to feel more productive and take fewer sick days, and that public image and the recruitment of and retention of employees are enhanced in green buildings.

Another approach to motivation distinguishes between "intrinsic" and "extrinsic" motivators. Intrinsic motivation is internalized and has to do with feelings of achievement, having a sense of control over events, and doing interesting and important work (Pink 2010). Extrinsic motivators come from outside the person and include incentives such as money, public recognition, or rewards. Behavioral change methods draw on both types of motivation.

Rewards and incentives, discussed earlier, may be the most commonly used types of motivators. On the one hand, in a workplace context rewards and incentives are commonplace and, like other motivators, could be applied to sustainability-related behaviors and outcomes. On the other hand, people and organizations have jobs to do and missions to achieve. Federal sustainability goals may not be viewed as integral to those jobs or missions and in some cases may be seen either as a luxury or as a distraction. Therefore, it may be important to appeal to multiple motivations that are relevant to the fuller context of the workplace, and not just to energy efficiency, resource use reduction, or greenhouse gas emission reductions. Developing an explicit connection to job and/or mission effectiveness may be important for individual and organizational motivation; for instance, energy is a strategic resource for the U.S. Department of Defense (DOD), and natural resource protection translates easily to sustainability in many agencies.

6.3. Achieving Broad and Persistent Impacts over Time

The methods and approaches discussed in this paper focus on structural changes in organizational and institutional practices and roles, as well as on deliberate efforts to change voluntary behaviors through psychosocial interventions. Although these methods are not widely used now in Federal or commercial buildings, there are pockets of experimentation occurring in the Federal sector. Translating these pockets of experimentation into ubiquitous, commonplace practice over time, within and across Federal agencies and to the private sector, involves several components.

Information sharing, through collaboration and interaction, is an important component of achieving sustainability within and across organizations — shifting it from an add-on program to how business is done. Collaboration among sustainability strategists is a means of identifying what practices and approaches do or do not work in particular settings (Laitner et al. 2009).

Collaboration can facilitate and amplify change as programs are implemented and evolve, equally useful to facility or energy managers as to organizational managers. Knowledge sharing among managers and staff can accelerate the pace of change through social engagement and "swarm intelligence."

Evaluation is essential for determining the extent to which strategies prove effective in the pursuit or achievement of stated goals. Interim evaluations during program implementation, which allow mid-course corrections, may be more important in the process of transforming organizations than outcome evaluations, which retrospectively gauge results achieved. Defining appropriate metrics to use in either form of evaluation is challenging for three main reasons. First, it may be difficult to define meaningful metrics that align with goals—items that lend themselves to quantification (e.g., number of pledges made) may not correlate well with desired results (e.g., number of pledges sustained). Second, it may not be possible to attribute impacts solely to a single behavioral intervention or change (e.g., energy saved solely because of pledges). Third, the metrics themselves may influence behavior, particularly if they are associated with incentives (rewards) or disincentives.

A third component is an agency or organizational narrative that connects the agencies' missions, responsibilities, productivity, and sustainability goals. A narrative, more than a simple logical argument, makes sustainability part of a larger story, rather than something separate, isolated, or even an impediment to the important work agencies conduct. The narrative provides a full rationale for change, metrics for change, and, perhaps, a sense of urgency for the need to change. Alternative future scenarios can indicate the implications of proceeding with the status quo versus undertaking the changes necessary to move toward sustainability.

7.0. Testing the Rules, Roles, and Tools Construct

A team of researchers from the Pacific Northwest National Laboratory tested the "rules, roles, and tools" construct in a project at Ft. Carson, a U.S. Army installation in Colorado (for more details, see Appendix B). The intervention focused on institutional and behavioral change to support the Army's Net Zero Energy goals for its military bases. Five buildings were selected for study. Two buildings were occupied by civilian personnel and three by military personnel.

The three-month intervention focused on enhancing the role of the Building Energy Monitor (BEM). At the time of the research, the BEMs had poorly defined roles and minimal training. The new training focused on how to conduct a walk-through energy use audit and how to support occupant behavior change. Pre- and post-intervention surveys of occupants addressed attitudes about energy use, comfort, and building perceptions.

The target behaviors were nighttime computer shutdown in all five buildings and temperature setbacks in two buildings with decentralized controls. Behavioral change methods included training BEMs on how to facilitate energy conservation behaviors (e.g., reminders to shut down computers and turn thermostats back at night, complimenting occupants when they performed the desired behaviors, addressing problems, providing reminders, and providing feedback). Bi-weekly emails, prepared by the research team for the BEMs, tracked how well each of the target buildings was performing as measured by percent of occupants complying with the computer shutdown. In addition, a letter to building occupants from Ft. Carson leadership supported the program. Results of the intervention showed high variation in compliance with the nighttime computer shutdown across buildings. Peak compliance rates ranged from 24 percent to 64 percent. The compliance rates were highest in the two civilian buildings, compared to the three buildings occupied by military personnel. While computer shutdowns resulted in a 1 percent electrical energy savings, the research team estimated that savings of 10 percent could be attained with 100 percent compliance. The nighttime thermostat setback resulted in 1 percent savings in natural gas.

Survey results showed an increase in the percent of occupants saying they had the skills to implement energy savings, felt personally responsible for saving energy in their buildings, and believed it was important.

The intervention also resulted in an unintended positive impact. Ft. Carson personnel are currently redefining and enhancing the role of BEMs to be more pro-active in achieving the Army's net zero energy, waste, and water goals.

8.0. Recommendations for an Action-Focused Framework

> Sustainability and resource conservation is a process. What happens on the ground will change over time as user behavior changes, space and building utilization shifts, and organization needs evolve.

This paper has described the integrated and interacting sets of behavioral issues that are key to achieving efficiencies of Federal and commercial buildings. It has shown that, at many levels, the people who own, manage, inhabit, and use buildings determine the extent to which sustainability and building performance goals are achieved.[6] It has highlighted the importance of institutional context and delineated the rules, roles, and tools components of that context. It has also described a variety of methods available to initiate and sustain lasting change, along with the evidence that supports their use in different circumstances.

Taken as a whole, this paper provides a high-level perspective on proven strategies and potential opportunities for Federal agencies and the private commercial building sector. Its "systems perspective" of occupant-use behaviors should enable broad organizational transformation in support of sustainability and reduced resource consumption that is significant, persistent over time, and able to be measured, evaluated, and documented. Following are high-level recommendations for implementing an action-based framework for changing individual and organizational use behaviors and achieving organizational sustainability and building performance goals:

1. Adopt a systems perspective to assess, design, implement, and evaluate sustainability and resource use strategies and programs as integrated technological and behavioral activities.

The tenets of behavior change in buildings encompass policymaking; regulation compliance, reinterpretation, or revision; operating procedures; infrastructural change; and individual actions. Behaviors that affect building performance are supported in part by specific building characteristics and requirements. Strategies and programs should modify applicable tenets, characteristics, and requirements in support of new behaviors that are consistent with sustainability objectives. Sustainable behaviors cultivated at the home or other non-workplace environments may also impact sustainability efforts in the workplace but must be researched further to better understand their potential.

[6] This goal achievement includes the entire building life-cycle from design, specification, construction, use, operation & maintenance, adaptation/reuse/renovation, and decommissioning and deconstruction/recycling.

2. Employ multiple methods and strategies that are supported by peer-reviewed scientific research as effective and aligned with specific sustainability goals and with mission fulfillment.

The methods and strategies supporting new behaviors serve different purposes and should be tailored to specific workplaces, contexts, and time frames. Employing multiple strategies instead of single, standalone strategies will better engage individuals and organizations possessing different motivations and responsibilities in the short and long terms.

3. Promote an enduring institutional transformation by evaluating outcomes to determine what strategies work best to build a culture of sustainability, what strategies are not effective and should be discontinued, and what strategies should be revised.

Behavioral and institutional changes are means for achieving sustainability goals, rather than ends in themselves. In creating a sustainability-oriented culture, people in their various organizational roles will expand on current activities and devise new strategies for conserving resources and improving the workplace sustainability goals.

4. Sponsor social science research that will build the evidence base for behavioral and institutional strategies aimed at saving energy, water, and materials, and at operating buildings sustainably over the long term.

This research should include carefully designed studies that incorporate monitoring and measuring, by collaborative teams of social scientists and building scientists, for 5–10 years, with periodic evaluations of the efficacy and persistence of institutionalized changes.

9.0. Research and Development Agenda for Institutional Change in Buildings

While there is a growing body of peer-reviewed literature on behavioral change, many significant gaps remain. In particular, the research on behavioral change related to sustainability has focused on individual behavior and primarily for the residential setting; it remains to be shown whether this knowledge is applicable wholly or in part to social action oriented toward energy and resource efficiency in agency workplace settings and across different subpopulations (e.g., building/energy managers, organizational managers, office workers). Complicating these questions are specific issues, such as the impacts of the absence of a commonly understood definition of sustainability or inadequate training and orientation resources, and the potential of alternative work arrangements (such as telework) or collaborations of smaller office-level groups to further sustainability goals. Moreover, much remains to be learned about organizational (versus individual), persistent (versus short-term), and sustainability-oriented change. Even the track record in the commercial sector, although growing, is not substantial — and the greater emphasis on profitability is a significant difference from government settings.

The number of case studies (both anecdotal and systematic) is growing; however, success stories often do not draw on scientific theory and methods to generate an improved evidence base. Thus, more rigorous analysis and social science expertise are needed to identify what does and does not work in different contexts. The actions Federal agencies have already undertaken to meet Executive Order 13514 (Federal Leadership In Environmental, Energy, and Economic Performance, from 2009) directives provide a natural experiment that is ripe for study and analysis.

The highest priorities for an R&D agenda, then, are to fill these kinds of gaps, such that findings can help Federal agencies take action to achieve significant, persistent, and measureable changes. In addition, an R&D agenda for this area should:

1. Integrate behavioral, institutional, and technological aspects of sustainability-related change.
2. Design projects to test social science findings (e.g., from behavioral economics or community-based social marketing) for their applicability in Federal settings and in sustainability efforts.
3. Use established theory and research design to implement projects that include:

    a. Rigorous baselining of comparable facilities and behaviors so that results can be compared
    b. Standard, well-defined methods allowing for scrutiny and replication
    c. Evaluation methods and dissemination approaches.

4. Develop reliable, robust measures and evaluation methods for behavioral and institutional change in Federal agencies — including established approaches such as interviews, focus groups, and surveys —

paying specific attention to sampling protocols, question design, and qualitative analysis that would yield comparable and reproducible data

Appendix A. Subcommittee on Buildings Technology Research and Development

A.1. BTRD Agency Members

- U.S. Department of Agriculture
- U.S. Department of Commerce
- U.S. Department of Defense
- U.S. Department of Energy
- U.S. Department of Health and Human Services
- U.S. Department of Homeland Security
- U.S. Department of Housing and Urban Development
- U.S. Department of Labor
- U.S. Department of Veterans Affairs
- U.S. Environmental Protection Agency
- U.S. General Services Administration
- National Aeronautics and Space Administration
- National Science Foundation
- Executive Office of the President
- Office of the Architect of the Capitol
- IDA Science and Technology Policy Institute
- Smithsonian Institution

Appendix B. Testing the "Rules, Roles, and Tools" Construct

A research team from the Pacific Northwest National Laboratory tested the "rules, roles and tools" construct at Fort Carson, an army installation in Colorado Springs in 2013. The project focused on institutional and behavioral change to support the Army's net zero energy goals for its military bases. A full version of the PNNL report can be found at: http://www.pnnl.gov/main/publications/external/technical_reports/PNNL-22824.pdf

B.1. Charting the Organizational Context

The first phase of the intervention identified the "rules, roles, and tools" that shape the context for energy conservation behaviors at Ft. Carson. This phase took about 40 percent of the project time line. The time commitment was necessary to identify personnel who could effectively support change, the specific behaviors to change, the systems and processes needed to support the change, and measurement systems needed to assess the impact of the intervention. This context evaluation also included a baseline survey, group discussions, and observations during site visits to help select target behaviors.

Through extensive discussions with Ft. Carson personnel, the team defined the key elements of the framework as follows:

- **Rules** –These rules included the formal policies and procedures related to energy use in the five demonstration buildings, as well as informal rules that might affect how a policy is perceived and the extent to which it is implemented. The importance of understanding both formal and informal rules is illustrated by the way occupants manage their temperature set points in one of the Ft.Carson buildings studied. While policies at Ft. Carson specify keeping thermostats at 70°F or below during heating season, practical thermostat and temperature calibration issues in one of the subject buildings meant that a thermostat setting of 75°F may have created a room temperature of only 68°F.
- **Roles** – Key individuals and groups who could make the most difference in achieving energy reduction goals were identified and engaged in the intervention design process. For this demonstration project, those roles included
    - occupants
    - building energy monitors (BEMs)
    - Ft. Carson senior leadership
    - computer network personnel.

BEMs are members of the building community who are responsible for monitoring building operations for energy-saving opportunities and reporting problems. The BEM program has been deployed Army-wide, although not all buildings at Ft. Carson have assigned BEMs.

- **Tools** – The research team reviewed existing technologies, systems, and processes and developed new tools that could be used to support the energy use reduction goals and help people do their jobs more easily and consistently. For example, the intervention leveraged the existing BEM to support occupant behavior change, but new tools including a floor-check form were developed by the research team to monitor key opportunities for improvement.

B.2. Intervention Methods

The intervention, which ran for three months, focused on two behaviors: (1) nighttime shutdown of desktop computers and (2) nighttime thermostat setbacks in two buildings with decentralized controls. These specific behaviors were selected because they could be easily performed by occupants and could result in a measurable impact on energy reduction.

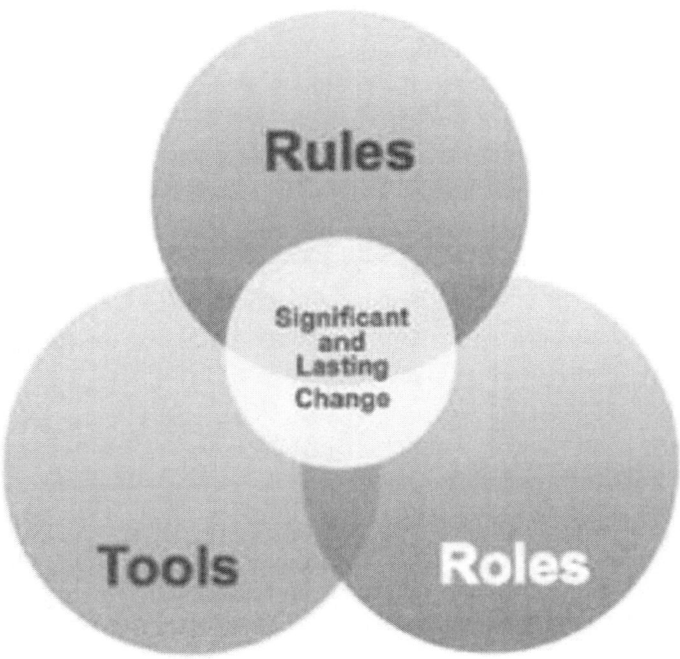

The behavior change methods focused on enhancing the role of the BEMs to help meet Ft. Carson's net zero energy goal. Training sessions focused on conducting weekly walk through checks of the buildings and acting as a "social facilitator" with occupants (e.g., reminding them to shut down computers and set thermostat backs at night, complimenting them when they performed the desired behaviors, providing reminders, and addressing problems). Bi-weekly emails to building occupants, prepared by the research team for the BEMS, discussed the progress of the intervention, identified how well each of the buildings was doing, and provided tips for improvement (e.g., setting up "reminders to self" to shut down computers at night and turn back the thermostat.) A final e-mail on week 8 of the intervention provided a graph of the compliance with energy shutdowns in each building as well as the overall results from the weekly checks.

In addition to enhancing the BEM role, the intervention also identified actions for Ft. Carson leadership in support of the project and working with computer network personnel to enable policy exemptions on nighttime computer shutdowns and to track compliance rates.

B.3. Results

B.3.1. Behavior and Attitude Change

Results of compliance with nighttime computer shutdown showed that peak compliance rates in each building ranged from 24 percent to 64 percent. The highest overall compliance rates were in the two buildings occupied by civilian personnel.

Post intervention survey results show a greater percent of occupants saying they had the skills to implement energy savings, felt personally responsible for reducing energy in their buildings, and believed it was important. Findings also support the idea that energy change efforts may be most effective if approached as a "high touch" activity through person-to-person communications. Specific factors that respondents said had the greatest influence on their energy conservation behavior were receiving:

- instruction from an immediate supervisor
- information on specific actions to take
- feedback on performance from a local advocate for energy conservation and,
- a letter from leadership encouraging occupants to take the specific conservation measures.

B.3.2. Energy Savings

The nighttime thermostat setbacks in the two buildings where this intervention was possible resulted in an estimated 1 percent energy savings in natural gas.

Calculation of electrical energy savings from the nighttime computer shutdowns showed a total of 1 percent savings during the intervention period. If compliance had been 100 percent, the energy savings could have been as high as 10 percent from this behavior change alone. The graph below shows the compliance rates for computer shutdown in each building over the 12 week intervention. Buildings 1118 and 1219 are occupied primarily by civilian personnel. The other three buildings are occupied by military personnel.

B.3.3. Conclusions and Lessons Learned

The research clearly shows that behavioral change interventions can result in energy savings. It also shows the importance of using the "rules, roles, and tools" construct through its emphasis on a critical-path role, that of BEMs. Enhancing the role of the BEMs and providing them with training on how to interact with occupants and to encourage behavioral changes to support Ft Carson's Net Zero Energy goal was key to successful outcomes.

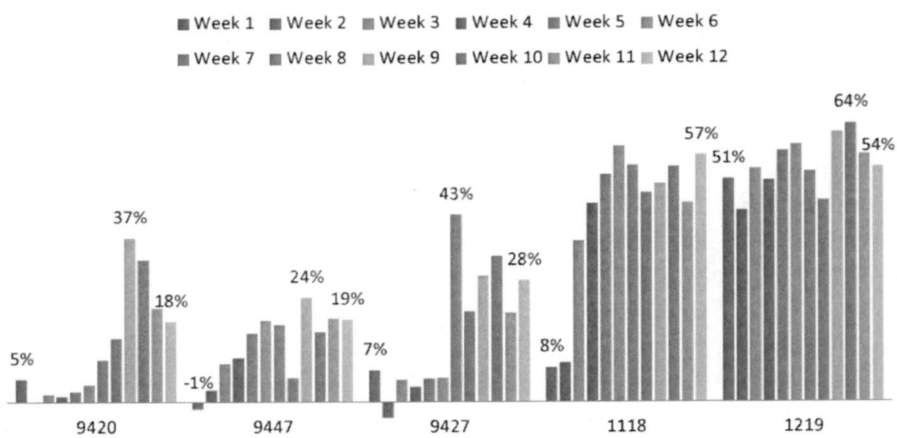

At the same time, this research project identified many of the challenges to behavioral change programs. Setting up the intervention was time intensive, particularly because the research team was not on site and needed to communicate largely by phone or email. Training BEMs remotely through a webinar limited the ability to engage in discussion, possibly decreasing the effectiveness of that training.

Despite the difficulties of putting the intervention in place and engaging with remote partners, the key indicator of the success of the rules, roles, and tools construct may be Ft. Carson's awareness that its facility management program needs to be transformed in the direction proposed by the research team. Ft. Carson currently is investigating how to implement the rules, roles, and tools construct across the installation.

APPENDIX C. REFERENCES

Abrahamse, W., L. Steg, C. Vlek, and T. Rottengatter. 2005. A review of intervention studies aimed at household energy conservation. *Journal of Environmental Psychology* 25, 273–291.

Aiken, Carolyn and Scott Keller 2009. The irrational side of change management. *McKinsey Quarterly* April 2009. Available at http://www.mckinsey.com/insights/organization/the_irrational_side_of_change_management

Ajzen, I. 1991. The theory of planned behavior. *Organizational Behavior and Human Decision Processes* 50, 179–211.

Argyris, C., and D. Schon. 1996. *Organizational Learning II*. Reading, Massachusetts: Addison Wesley.

Ariely, D. 2009. *Predictably Irrational: the Hidden Forces that Shape Our Decisions*. New York: Harper.

Armel, C. 2008. Energy, behavior, and design. October 28. Precourt Institute for Energy Efficiency and Stanford University.

Aune, M., T. Berker, and R. Bye, 2009. The missing link which was already there: building operators and energy management in non-residential buildings. *Facilities, Vol 27 (*1/2), 44-55.

Becker, L.J. 1978. Joint effect of feedback and goal setting on performance: a field study of residential energy conservation. *Journal of Applied Psychology* 63(4), 428–433.

Buildings Technology Research and Development (BTRD), Subcommittee on. 2011. *Submetering of Building Energy and Water Usage: Analysis and Recommendations of the Subcommittee on Buildings Technology Research and Development.* Washington, DC: National Science and Technology Council Committee on Technology. http://www.whitehouse.gov/sites /default/files/microsites/ostp/submetering_of_building_energy_and_wate r usage.pdf.

Campbell, M., D. Buckeridge, J. Dwyer, S. Fong, V. Mann, O. Sanchez-Sweatman, A. Stevens, and L. Fung. 2000. A systematic review of the effectiveness of environmental awareness interventions. *Canadian Journal of Public Health* 91(2), 137–143.

Carrico, A.R., and M. Riemer. 2011. Motivating energy conservation in the workplace: an evaluation of the use of group level feedback and peer review. *Journal of Environmental Psychology* 31, 1–13.

Cialdini, R.B. 2010. *Influencing Change: Applying Behavioral Science Research Insights to Reframe Environmental Policy and Programs.* Sacramento, CA: Behavior, Energy and Climate Change Conference.

Cialdini, R.B. 2007.*Influence: The Psychology of Persuasion*, 3rd Ed. New York: Collins Business.

Cialdini, R.B. 2005. Basic social influence is underestimated. *Psychological Inquiry* 16(4), 158–161.

Consolvo, S., D.W. McDonald, and J.A. Landay. 2009. *Theory-Driven Design Strategies for Technologies that Support Behavior Change in Everyday Life*. Boston, MA: CHI 2009.

Cole, R.J. 2011. Motivating stakeholders to deliver environmental change. *Building Research & Information* 39,431–435.

Costanzo, M., D. Archer, E. Aronson, and T. Pettigrew. 1986. The difficult path from information to action. *American Psychologist* 41(5), 521–528.

Darby, S. 2006. *The Effectiveness of Feedback on Energy Consumption: A Review for DEFRA of the Literature on Metering, Billing and Direct Displays*. Oxford, England: Environmental Change Institute, University of Oxford.

Dietz, T., G.T. Gardner, J. Gilligan, P.C. Stern, and M.P. Vandenbergh. 2009. Household action can provide a behavioral wedge to rapidly reduce U.S. carbon emissions. *Proceedings of the National Academy of Sciences* 106, 18452–18456.

Duhigg, Charles 2012. *The Power of Habit: Why We Do What We Do in Life and Business*. Random House, New York.

Ehrhardt-Martinez, K., and J.A. Laitner. 2010. *Behavior, Energy, and Climate Change: Emerging Policy Directions, Program Innovations, and Research Paths*. Washington, DC: U.S. Department of Energy and Argonne National Laboratory.

Energy Independence and Security Act (EISA 2007). Public Law 110–140. Energy Policy Act of 2005 (EPAct 2005). Public Law 109–158.

EPRI. 2009. *Residential Electricity Use Feedback: A Research Synthesis and Economic Framework*. Palo Alto, CA: EPRI.

Erhardt-Martinez, K. 2010. *Changing Habits, Lifestyles, and Choices: the Behaviors that Drive Feedback-induced Energy Savings.* Asilomar, CA: American Council for an Energy Efficient Economy Summer Study.

Executive Order [E.O.] 13514. October 8, 2009. Federal Leadership in Environmental, Energy, and Economic Performance. *Federal Register* 74(194), 52117–52127.

Foster, B., and S. Mazur-Stommen. 2012. *Results from Recent Real-Time Feedback Studies.* American Council for an Energy-Efficient Economy (ACEEE) http://aceee.org/research-report/b122.

Froelich, J. 2009. *Promoting Energy Efficient Behaviors in the Home through Feedback: The Role of Human-Computer Interaction.* White paper. Seattle, WA: DUB Institute, University of Washington.

Froelich, J., L. Findlater, and J. Landay. 2010. The design of eco-feedback technology. *Proceedings of the 28th International Conference on Human factors in Computing Systems.* Atlanta, GA: CHI 2010.

Gardini, J., G. Giulliani, and M. Marricchi. 2011. Finding the right place to start change. *McKinsey Quarterly.* http://www.mckinseyquarterly.com /Finding_the_right_place_to_start_change_2890.

Gatersleben, B., and C. Vlek 1998. Household consumption, quality of life and environmental impacts. In K.J. Noorman, and A.J.M. Schoot-Uiterkamp (eds.), *Green Households? Domestic Consumers, Environment and Sustainability.* London: Earthscan, 141–183.

Gershon, D. 2009. *Social Change 2.0: A Blueprint for Reinventing the World.* High Point, NC: Chelsea Green.

Goins, J. and M. Moezzi 2011. *Case study of CalSTRS Headquarters.* Center for the Built Environment, Berkeley, CA. http://www.cbe.berkeley.edu /research/pdf_files/calstrs-final-2011.pdf.

Goins, John and Mithra Moezzi 2013. Linking building performance to occupant complaints. *Building Research & Information* 41(3), 461-372.

Goldstein, N.J., R.B. Cialdini, and V. Griskevicius. 2008. A room with a viewpoint: using social norms to motivate environmental conservation in hotels. *Journal of Consumer Research* 35 (electronic version).

Griskevicius, V., R. B. Cialdini, N. J. Goldstein. 2008. Applying (and resisting) peer influence. *MIT Sloan Management Review* (49)2, 84–88.

Haanaes, K., M. Reeves, I. von Streng Velken, M. Audretsch, D. Kiron, and N. Kruschwitz. 2012. *Sustainability Nears a Tipping Point: Findings from the 2011 Sustainability & Innovation Global Executive Study and Research Project.* MIT Sloan Management Review Special Report. http://sloanreview.mit.edu/feature/sustainability-strategy/.

Hauge, A.L., J. Thomsen and T. Berker 2010. User evaluations of energy efficient buildings— literature review and further research. In Zero Emission Buildings—Proceedings of Renewable Energy Conference 2010, Trondheim,Norway.

He, H.A., S. Greenberg, and E.M. Huang. 2010. *One Size Does Not Fit All: Applying the Transtheoretical Model to Energy Feedback Technology Design*. Atlanta, GA: CHI 2010.

Heath, D., and C. Heath. 2010. *Switch: How to Change Things When Change Is Hard*. New York: Crown Publishing House.

Heffner, R.R., T.S. Turrentine, and K.S. Kurani. 2006. *A Primer on Automobile Semiotics*. Davis, CA: Institute of Transportation Studies, University of California.

Janda, Kathryn B. 2014. Building communities and social potential: Between and beyond organizations and individuals in commercial properties. *Energy Policy* 67:48-55.

Katzev, R.D., and T.R. Johnson. 1984. Comparing the effects of monetary incentives and foot-in-the-door strategies in promoting residential electricity conservation. *Journal of Applied Social Psychology* (14)1, 12–27.

Kahneman, D. 2011. *Thinking, Fast and Slow*. New York: Farrar, Straus and Giroux.

Kotter, J. 1996. *Leading Change*. Boston: Harvard Business School Publishing Corporation.

Laitner, J.A., K. Ehrhardt-Martinez, and C.P. Knight. 2009. *The Climate Imperative and Innovation Behavior: Encouraging Greater Advances in the Production of Energy Efficient Technology and Services*. Berkeley, CA: California Institute for Energy and Environment.

Lenoir, Aurélie , Shaan Cory, Michael Donn and François Garde 2011. Users' behavior and energy performances of net zero energy buildings. In Proceedings of the 12th Conference of International Building Performance Simulation Association, 14–16 November 2011, Sydney, Australia. http://www.ibpsa.org/proceedings/BS2011/P_1511.pdf

Lewin, K. 1951. *Field Theory in Social Science: Selected Theoretical Papers*. D. Cartwright (ed.). New York: Harper & Row.

Lobato, C., S. Pless, M. Sheppy, and P. Torcellini. 2011. *Reducing Plug and Process Loads for a Large Scale Low Energy Office Building: NREL's Research Support Facility*. Las Vegas, NV: ASHRAE Winter Conference.

McKenzie-Mohr, D. 2009. *Fostering Sustainable Behavior*. Washington, DC: Behavior, Energy, and Climate Change Conference.

McKenzie-Mohr, D., N.R. Lee, P.W. Schultz, and P. Kotter. 2012. *Social Marketing to Protect the Environment*. Los Angeles: Sage.

McKenzie-Mohr, D., and W. Smith. 1999. *Fostering Sustainable Behavior*. British Columbia, Canada: New Society Publishers.

McKenzie-Mohr, D., and W. Smith. 2007. *Fostering Sustainable Behavior: An Introduction to Community-Based Social Marketing*. Gabriola Island, British Columbia: New Society Publishers.

Metzger, I., A. Kandt, and O. VanGeet. 2011. *Plug Load Behavioral Change Demonstration Project*. National Renewable Energy Laboratory, Golden, CO, NREL/TP-7A40-52248 http://www.nrel.gov/docs/fy11osti/52248.pdf.

Miller, N.G., and D. Pogue. 2009. *Do Green Buildings Make Dollars and Sense?* Burnham-Moores Center for Real Estate, University of San Diego; and CB Richard Ellis.

Moezzi, M. 2009. *Social Sciences for Reduced Energy Use in Buildings: Past Contributions and Future Potential*. Washington, DC: Integrating Social and Behavioral Insights to DOE Programs and Policies.

Moezzi, Mithra and Kathryn B. Janda 2013. Redirecting research about energy and people: from "if only" to "social potential." ECEEE Summer Study Proceedings. ("Middle-out")

Moezzi, M., M. Iyer, L. Lutzenhiser, and J. Woods. 2009. *Behavioral Assumptions in Energy Efficiency Potential Studies*. Oakland, CA: California Institute for Energy and Environment.

Nakajima, T., V. Lehdonvirta, E. Tokunaga, and H. Kimura. 2008. *Reflecting Human Behavior to Motivate Desirable Lifestyles*. Washington, DC: DIS 2008.

Nicoll, G. 2006. *Taking the Stairs: Environmental Features that Explain Why People Use Stairs in 3 to 4 Academic Workplace Buildings*. PhD Thesis, Georgia Institute of Technology, Atlanta.

Pallak, M.S., and N. Cummings. 1976. Commitment and voluntary energy conservation. *Personality and Social Psychology Bulletin* 2(1), 27–31.

Pasquier, S.B. 2011. *Saving Electricity in a Hurry: 2011 Update*. Paris, France: International Energy Agency. http://www.iea.org/papers/2011/saving_electricity.pdf, accessed 2/23/12.

Pickens, P.M. 2002. *Community Based Social Marketing as a Planning Tool*. Masters Thesis, University of Oregon.

Pink, D. 2010. *Drive: The Surprising Truth about What Motivates Us*. New York: Riverhead Books.

Power System Engineering. 2010. *Measurement and Verification Report of OPower Energy Efficiency Pilot Program.* Madison, Wisconsin: Power System Engineering.

Prindle, W.R. 2010. *From Shop Floor to Top Floor: Best Business Practices in Energy Efficiency.* Washington, DC: Pew Center on Global Climate Change.

Prochaska, J.O., and W.F. Velicer. 1997. The transtheoretical model of health behavior change. *American Journal of Health Promotion* 12(1), 38–48.

Rogers, E. 1995. *Diffusion of Innovation*, 5th Ed. New York, NY: Free Press.

Schein, E.H. 1999. *The Corporate Culture Survival Guide: Sense and Nonsense about Cultural Change.* San Francisco: Jossey-Bass.

Schultz, P.W., J.M. Nolan, R.B. Cialdini, N.J. Goldstein, and V. Griskevicius. 2007. The constructive, destructive and reconstructive power of social norms. *Psychological Science* 18(5), 429–434.

Shui, Bin 2012. *Greening Work Styles: An Analysis of Energy Behavior Programs in the Workplace.* American Council for an Energy-Efficient Economy, Washington, DC.

Siero, F.W., A.B. Bakker, G.B. Dekker, and M.T.C. Van Den Burg. 1996. Changing organizational energy consumption behavior through comparative feedback. *Journal of Environmental Psychology* 16, 235–246.

Smith, J. 2005. *Obtaining Behavior Change Not Just Raising Awareness.* Adelaide, Australia: Australian Association of Environmental Education, SA Division Workshop. http://www.emergingself.com.au/About_Jodi-Anne.htm.

Stern, P.C. 2002. Changing behavior in households and communities: what have we learned? In T. Dietz, and P.C. Stern (eds.), *New Tools for Environmental Protection.* Washington, D.C.: National Academy Press.

Stern, P. 2009. *Designing Effective Policies Using Social Science Insights.* Washington, DC: DOE Energy and Behavior Workshop.

Thaler, R.H., and C.R. Sunstein. 2009. *Nudge: Improving Decisions about Health, Wealth and Happiness.* New York: Penguin.

Triandis, H. 1977. *Interpersonal Behaviour.* Monterey, CA: Brooks/Cole.

Turner, C., and M. Frankel. 2008. *Energy Performance of LEED for New Construction Buildings.* Vancouver, WA: New Buildings Institute. www.usgbc.org/ShowFile.aspx?DocumentID=3930.

Tversky, A., and D. Kahneman. 1974. Judgment under uncertainty: heuristics and biases. *Science* 185, 1124–1131.

United States Department of Energy (DOE). 2009. *Big Results, Bigger Potential: CFL Market Profile.* Washington, DC: DOE. http://www.energystar.gov/ia/products/downloads/CFL_Market_Profile.pdf.

van Houwelingen, J.H., and W.F. van Raaij. 1989. The effect of goal-setting and daily electronic feedback on in-home energy use. *Journal of Consumer Research* 16(1), 98–105.

Verplanken, B., and W. Wood. 2006. Interventions to break and create consumer habits. *Journal of Public Policy and Marketing* 25(1), 90–103.

Weber, E. U. & Lindemann, P. G. 2007. From intuition to analysis: Making decisions with our head, our heart, or by the book, appearing in H. Plessner, C. Betsch & T. Betsch (Eds.), Intuition in judgment and decision making, 191-208. Mahwah, NJ: Lawrence Erlbaum.

Winkler, R.C., and R.A. Winett. 1982. Behavioral interventions in resource management: a systems approach based on behavioral economics. *American Psychologist* 37, 421–435.

Wood, W., and D.T. Neal. 2009. The habitual consumer. *Journal of Consumer Psychology* 19(4), 579– 592.

Zhang, H., E. Arens, D.E. Kim, E. Buchberger, F. Bauman, and C. Huizenga. 2009. Comfort, perceived air quality, and work performance in a lower power task-ambient conditioning system. *Building and Environment* 45, 29–39.

In: Buildings, Resource Consumption … ISBN: 978-1-63482-767-6
Editor: Joanna Cruz

Chapter 2

SUBMETERING OF BUILDING ENERGY AND WATER USAGE: ANALYSIS AND RECOMMENDATIONS OF THE SUBCOMMITTEE ON BUILDINGS TECHNOLOGY RESEARCH AND DEVELOPMENT*

Buildings Technology Research and Development Subcommittee

EXECUTIVE OFFICE OF THE PRESIDENT
NATIONAL SCIENCE AND TECHNOLOGY COUNCIL
WASHINGTON, D.C. 20502

October 21, 2011

* This is an edited, reformatted and augmented version of a report issued by the Executive Office of the President, National Science and Technology Council, October 2011.

Dear Colleague:

I am pleased to forward this report entitled *Submetering of Building Energy and Water Usage*, which provides an overview of the benefits and complexities of advanced metering technologies for fine-grained and real-time measurement of energy and water systems in buildings. The use of advanced metering technologies supports resource-efficient design and retrofit strategies that can drive energy and water consumption reduction. Commercial and residential buildings consume about one-third of the world's energy and account for more than 40 % of total U.S. energy consumption. If current trends continue, by 2025 buildings worldwide will consume more energy than the transportation and industry sectors combined.

The refined measurement of energy and water use represents a key enabler for the improved performance of new and existing buildings. For building operators, a detailed record of system performance provides a critical means of not only detecting system malfunctions easily but also focusing future design and retrofit activities on the most cost-effective energy and water system improvements. For building occupants, detailed information on consumption promotes resource conservation through behavioral changes.

The Federal sector has already demonstrated leadership in conserving energy and water resources through the introduction of advanced metering. I anticipate that this report will provide the basis to initiate new Federal submetering demonstrations that can document even greater energy- and water-use effectiveness that will in turn result in increased adoption of submetering by Federal and private-sector building owners.

Sincerely,

John P. Holdren
Assistant to the President for Science and Technology
Director, Office of Science and Technology Policy

EXECUTIVE SUMMARY

The submetering of buildings, through the refined measurement of energy and water use, enables the improved performance of new and existing buildings. Submetering provides the operations and maintenance transparency necessary to enable more efficient management of energy and water resources. In addition, submetering can drive behavioral change related to energy conservation and advance real-time building interaction with the Smart Grid. Each of these potential benefits can dramatically improve building performance and lead to reduced resource consumption. For building operators, a detailed record of system performance data is critical not only to detect system malfunctions, but also to focus future design and retrofit activities on the most cost-effective energy and water system improvements. For building occupants, submetering provides detailed information on their consumption behaviors, which in turn promotes resource conservation.

This report provides an overview of the key elements of submetering and associated energy management systems to foster an understanding of the many potential benefits and complexities associated with use of these systems. While submeters by themselves have no direct impact on resource use, the data they capture informs real-time energy and water performance, can pinpoint performance variations over time or relative to other buildings, feeds into building automation systems that drive continuous operational improvements, and provides the information needed to encourage behavioral and operational changes by building operators and occupants.

This report lays the groundwork for the economic and operational benefits achievable through the sub-metering of buildings. Evaluating the feasibility and impact of a submetering investment requires a life cycle cost approach to accurately assess the long-term economic costs and benefits of implementation. Costs include not only hardware and software, but also the regular maintenance of submetering systems by properly trained O&M staff and an ongoing program of behavioral and operational improvements for building operators and occupants.

The benefits of submetering vary by building type, use, and mode of operation, as well as by external factors such as climatic zones, utility regulations and policies, and energy and water rate structures. The case studies presented herein offer supporting evidence that the resource savings and economic return on investment in submetering can be substantial. They also highlight benefits that can emerge when submetering is integrated with building automation systems and with operations and maintenance (O&M) procedures. Equally important are technical details of submeters and the accompanying data networks, software, and services that gather and synthesize meaningful information for building operators and occupants. The manner and mode of this acquisition and presentation of information influence the rate of acceptance by these same stakeholders and the gains enabled by submetering.

Ongoing behavioral studies focused on energy conservation for buildings yield a good understanding of how human behaviors affect energy conservation, especially in the residential sector. For commercial buildings, Vornado Realty Trust estimated that between 60 and 80 percent of the energy used in commercial office buildings is consumed by tenants within their spaces, yet no widely available tools exist to help tenants understand their energy consumption or to compare it against their peer group.[1] Analogously, O&M staff lacking the proper tools and training may overlook opportunities to improve building operations enabled by submetering. In both cases, appropriate instruction, feedback mechanisms, and incentives are central to encouraging the changes that result in energy-efficient building operations and energy conservation using submetering capabilities.

The building community should view submetering as an essential component of future building operational improvements for energy efficiency and conservation improvements, to monitor-based continuous commissioning, and to improve indoor environmental quality. Submetering systems must be carefully designed to meet stated operational criteria and objectives, i.e., data analysis and operations management requirements must guide submetering hardware and software selections and specifications of the system configuration. In addition, submetering systems must be installed and maintained by properly trained and incentivized O&M technical professionals to achieve the design performance objectives.

The BTRD Subcommittee offers the following recommendations to building program managers, owners and operators, as well as to Federal and state policy officials in their examination of submetering as a potential tool to meet energy and water conservation goals. These recommendations cover a range of issues, both human and operational—from design to procurement, installation, and ongoing full production operations.

Recommendation 1: As part of overall building design and retrofit projects, building owners should evaluate the economic and technical feasibility of submetering using life cycle cost analysis for the economic justification and business-case development.

Recommendation 2: When evaluating submetering opportunities, design teams should collaborate with building owners, operators, and occupants, and design, engineering, and information technology professionals to identify data needs and appropriate submetering systems and configurations.

Recommendation 3: Submetering systems should, when possible, be integrated into building automation systems and used to identify building system impairments or dysfunction, to improve operational procedures, drive behavioral changes that improve building energy efficiency and conservation, and augment building-side automation systems that interface with the Smart Grid.

Recommendation 4: Submetering information and feedback mechanisms should be tailored to intended users to effect operational and behavioral change in building operators and occupants. Information and feedback should be of an appropriate form and frequency and should focus on useful breakdowns of resource use; include historical data when appropriate to facilitate comparisons with past use; and, when possible, suggest specific actions that can be taken to reduce consumption.

Recommendation 5: Submetering systems should be operated as production-quality building management systems with high levels of reliability, information assurance and security, established data acquisition/ archival procedures, and periodic system maintenance requirements.

1. Introduction

1.1. Submetering and the Smart Grid

With Federal funds from the American Recovery and Reinvestment Act (ARRA) apportioned for weatherization programs, loan guarantees, block grants, and new investments in green technology, initiatives promoting resource efficiency and the use of renewable resources have begun to make significant progress. As government and private institutions invest time and resources, proper prioritization of building projects and initiatives becomes essential to responsible fiscal management. To properly direct these efforts, planners must understand where energy or water is consumed, at what rate and cost, and how current consumption compares to historical baselines or similar buildings. Historical and comparative analysis provides context for resource consumption and can identify areas for improved efficiency. By narrowing the scope of the examination to specific buildings, rooms, areas, or systems, this knowledge helps to pinpoint potential resource savings. When used for this purpose, sensing technologies go beyond their passive measurement role to active and continuous remediation, resulting in significant resource savings and improved overall building operation and maintenance.

Meters are simply sensors that measure and record resource use, such as energy or water consumption. Metering can be employed at several different scales, and is used to support effective management of a building's systems, including management of comfort, energy and water consumption, delineated cost burdens, and investment decisions in the short and long term. Historically, utility metering has occurred on an aggregate basis, with measurements taken for whole buildings or campuses roughly once per month. While this model has supported utility cost-recovery and billing practices, in recent years the utility sector has looked to greater integration of information technologies as potential sources of resource conservation and economic savings. Associated technologies promise a variety of benefits through more accurate and timely management of system operations, as well as a greater sharing of resource consumption information with consumers.

As a prominent example, electrical-energy stakeholders have presented alternate visions of the future smart grid, with an enhanced information and communications infrastructure embedded into the electrical distribution system. Funds from the American Recovery and Reinvestment Act (ARRA) legislation have enabled significant investments in an advanced electrical metering infrastructure, which allows for more frequent data collection on electricity use and other power characteristics.

Thus far, among consumer-focused smart grid technologies, smart meters have received the bulk of public attention and funding, with less emphasis placed on submeters or the metering of resources not related to energy. Whereas advanced meters generate more temporally resolved longitudinal data, submeters provide more spatially resolved measurements for individual areas, systems, or equipment. Moreover, while utility companies have led smart meter deployments, submetering requires involvement of and investment by individual building owners as well as more detailed tailoring to specific building configurations.

1.2. Fundamentals of Submetering

Resource use can be measured from an entire campus down to an individual building, from a building system or interior space, or even at the scale of an occupant's end-use load. As metering infrastructures are deployed at greater densities, more spatially and temporally resolved data become available; advanced metering may provide data at hour- or minute-long intervals at a full building scale, while submetering is capable of providing data at near-continuous time resolution and at a sub-building scale. As an example, end-use loads from electrical devices used by building occupants (both commercial and residential) represent the area of largest demand growth in buildings.[2] Through the use of submeters, these plug loads may be quantified using an individual measuring device at each plug outlet or at multiple outlets along a single distribution line.

Higher-resolution data points are necessary for certain types of resource reduction strategies. In general, there are two ways a building manager can pursue reductions in energy or water use: through increased resource efficiency or through resource conservation. Resource efficiency is usually increased through the installation of new technologies that deliver the same service (lighting, heating and cooling, hand washing, etc.) while using less energy or water. Resource conservation, on the other hand, is achieved through improved operation and maintenance (O&M) practices and/or occupant behaviors such as reducing unnecessary lighting or heating loads and reprogramming energy control systems.

Metrics for both efficiency and conservation require knowledge of current resource use and practices, generally at the sub-building level, but the temporal scales for the data may be different. Retrofits for resource efficiency can sometimes be identified through snapshot data such as those obtained through an energy audit. While helpful, energy audits are of limited utility in that they provide data only for one specific period of time. Submetering, on the other hand, can drive long-term and deeper efficiency and conservation improvements by introducing a localized, continuous timescale. This refined timescale provides insight into daily, weekly, or seasonal O&M issues, occupant behaviors, performance of installed equipment (e.g., HVAC and lighting), and verification of installed efficiency technologies. A time-series approach to analysis also provides the necessary insight to drive conservation through changes in occupant behaviors or continual improvements to building O&M procedures as conditions change over time. Moreover, these refined data points may reveal other system problems or potential efficiency measures that were not immediately apparent with snapshot data.

These main benefits—identifying operation improvements and efficiency retrofit opportunities—and others are discussed in Section 2 along with case studies illustrating real-world examples of their implementation. Section 3 elaborates on the economic calculations needed to evaluate the costs and benefits of submetering. To be useful in decision-support, information from meters, submeters, and other sensors must flow to the relevant entities making operational, behavioral, and investment decisions—building occupants, facilities managers, metering data management companies, and in some cases, utility companies.

However, while data from submeters and other sensors can increase awareness of energy consumption and motivation to reduce it, actually changing a building's consumption profile and achieving expected financial payback from the submetering investment may require new building management practices, financial incentives, enabling technologies, decision-support tools, and input from operations and human factors research. These requirements can only be met through the integration of submetering data with existing building automation systems, discussed in Section 4. As submeters and other sensors are added to a building and connected to building automation systems, the data points are transmitted via communications networks and analyzed for use by O&M staff and occupants, as well as by facilities managers, possibly in remote locations. Such coordination is especially important in complex commercial and multi-unit residential buildings, in which many systems overlap, and decisions affecting operating characteristics (e.g., how air conditioning needs are supplied) may be shared among multiple parties including occupants, owners, building managers, and utilities. These sensing and control technologies separate the decision-making into smaller elements, with oversight and management offered to those most appropriate to the particular task, such as the room occupant(s), the facility manager, or the smart grid's operations manager.

In many situations, these operational decisions can be partially or wholly automated; common examples include: motion sensor-activated light switches, elevator scheduling, and temperature controls. Improvements in all of these areas—from sensing and mechanical controls to analytics and human interface design—push toward the "intelligent building" paradigm, where appropriate levels of automation are used for the task and the occupants' needs, and building systems and managers are more responsive and adaptive to consumption factors such as weather and occupants' use of space.

1.3. Regulatory Context

Until recently, despite the potential benefits of detailed resource measurement in buildings, there have been relatively few instances where either building-level metering or submetering was specifically required by Federal or local regulations. Appendix A discusses in more detail the relevant Federal, state, and local building code regulations. A summary of these requirements is provided in this section.

In the Federal sector, building-level metering of electricity is required by Section 103 of the Energy Policy Act of 2005 (EPAct 2005). The Energy Independence and Security Act of 2007 (EISA 2007) built on this precedent by requiring that agencies establish energy benchmarks for their facilities' portfolios, implement life cycle cost-effective energy management projects through both commissioning and retrofits, and monitor and verify performance of the measures they implement. The Department of Defense (DoD) Instruction 4170.11, Installation Energy Management (November 22, 2005), requires DoD to meter electricity, natural gas, and water at appropriate facilities and to meter steam at steam plants. Executive Order 13514, signed in 2009 by President Obama, expanded the previous energy and water goals to include more comprehensive sustainability goals for the Federal government, such as reductions in greenhouse gas emissions and in water consumption.

Advanced metering (including submetering systems) is a critical catalyst for compliance with these requirements because it enables identification of life cycle cost-effective measures, provides benchmark data, and supports monitoring and verification of energy performance. The Department of Energy's Federal Energy Management Program (FEMP) has issued guidance for this requirement in its Guidance for Electric Metering in Federal Buildings,[3] and in its best practices guide entitled Metering Best Practices: A Guide to Achieving Utility Resource Efficiency.[4]

In the private sector, building-energy use is regulated mainly at the state and local levels through building energy design codes. For commercial codes, the ANSI/ASHRAE/IESNA Standard 90.1—Energy Efficient Design of New Buildings Except Low-Rise Residential Buildings—provides the basic compliance standard for new construction in most jurisdictions. Although the 1989 version of this code briefly required certain types of submetering—separate metering for tenants in multi-tenant buildings and for shared electrical, heating/ventilation/air conditioning (HVAC) or service water heating (SWH) systems—these provisions were eliminated from all subsequent versions, including the most recent 2010 edition. As a design standard, submetering requirements proved difficult to justify solely on a cost basis when their benefits would not be assessed until after the building was occupied and if the metering data points were used in ways that resulted in energy conservation or energy efficiency improvements through other activities.

Recently, there has also been action at state and local levels. For instance, the New York City Council passed legislation requiring the installation, by January 1, 2025, of submeters in tenant spaces that (i) exceed 10,000 gross square feet on one or more floors of a Covered Building let or sublet to the same person, or (ii) consist of a floor of a Covered Building larger than 10,000 gross square feet consisting of tenant spaces let or sublet to two or more different persons. Once installed, the meters must be read at least on a monthly basis, and each tenant must be provided with a monthly statement showing the amount of electricity measured for each area covered by the submeter. This requirement is just one of many approved to mandate ongoing efficiency improvements in large existing commercial buildings.

In the absence of specific regulatory requirements, submetering decisions often rely upon economic savings and payback justification. The Federal government's efficiency retrofit policies require valuation methodologies such as savings-to-investment ratios, payback periods, and life cycle costs (on a 40-year basis) to ensure these retrofits provide value. The appropriate economic analysis methods commonly used are discussed further in Section 3.

Submetering systems are unusual among building retrofits in that they themselves do not reduce resource consumption; rather, they provide the sensor infrastructure and data necessary for the efficient operation of buildings and for accurate benchmarking of performance. This distinction presents the challenge of associating a direct and clear means of linking specific energy cost reductions with the submetering investment, handicapping these technologies vis-à-vis other investments. However, submetering systems may also provide indirect benefits such as ongoing and improved operational efficiencies, resulting in lower building-wide O&M costs and reduced energy-related emissions, all of which may be difficult to quantify in purely financial terms.

2. BENEFITS AND USES OF SUBMETERING DATA

Submetering can benefit building owners, O&M staff, and business operations managers in several ways. As discussed above, buildings may be submetered to measure resource use of heating, cooling, or lighting systems and reduce these loads in response to market drivers (e.g., utility costs) or regulation. Individual systems or pieces of equipment may be submetered to determine if they are "working efficiently" with regard to both economic and comfort considerations. There are also practical benefits in some cases—a building or campus currently served by one utility-owned master meter may be submetered for different tenants to promote more appropriate allocation of utility charges and to effect potential reductions in peak demand costs. Finally, submetered data can be used to provide feedback on energy consumption to tenants or building users to promote behavioral change that leads to energy conservation. These different uses of submetering may also be combined within a single building. For example, a building on campus served by a single utility meter might be submetered to facilitate more equitable billing of tenants as well as to better manage particular systems, subsystems, or end-use plug loads.

All of these forms of submetering are useful, as they provide building owners with the measurement capability to better control resource use in buildings. It is imperative that the efficient control of energy and water usage be exercised in a manner that minimizes energy and water consumption while properly maintaining a building's indoor environment and functionality. It is also the case that submetering data analysis can identify situations where energy or water operations and management can be modified to provide a higher quality environment and better meet occupant needs.

The following text box lists many commonly cited submetering benefits in four categories. The first, enabling monitoring-based commissioning, involves tracking the performance of building equipment over time and optimizing it to save energy and water and to increase user comfort. The second, identifying and monitoring building retrofits, involves the use of submeters to both locate where retrofits will save the most resource and to verify that these planned savings actually occur. The third, aligning incentives for behavioral change, involves providing either direct monetary (i.e., tenant-specific billing) or non-monetary (i.e., energy use compared to peers) feedback to induce resource conservation behavior among building users. Finally, an ancillary benefit of submetering may be a reduction in peak demand charges via utility bills, provided that the utility accepts "virtual aggregation" as an acceptable billing scheme. These benefits are discussed further in the next sections, complete with case studies describing their application.

Case Study Descriptions

1) Enabling Monitoring-Based Commissioning

- Identifying performance improvements and guiding preventive maintenance—trends in monthly and annual use of water and energy help to identify the benefits of system upgrades and the systems (e.g., boilers, cooling towers) that may need attention following unexpected increases in consumption.
- Enabling quick response to failures of system components, assuming the meters are linked to an energy management system (EMS) or a building management system (BMS).
- Supporting life cycle financial planning by developing a dataset to support trend analysis.

- Focusing accountability for building operations on the facilities department, encouraging building managers to control energy and water consumption.

2) Identifying and Monitoring Efficiency Retrofits

- Verifying savings from energy and water improvement projects.
- Helping to compile baseline energy use for setting contractual terms with an energy service company (ESCO).
- Helping to make decisions about energy and water upgrades in buildings by comparing use in similar facilities.

3) Aligning Incentives and Enabling Behavioral Conservation

- Facilitating charge-backs to departments or other campus units as a way to encourage energy and water efficiency measures.
- Providing data to building occupants about the impacts of their behavior on energy and water consumption.

4) Peak Demand Reductions

- Lowering peak demand charges on electrical utility bills through virtual aggregation of different submeters.

Taken in part from ENERGY STAR®, "Sub-Metering Energy Use in Colleges and Universities: Incentives and Challenges," http://www.energystar.gov/ia/business/higher_ed/Submeter_energy_use.pdf.

Reducing overall energy consumption through efficiencies gained by submetering can provide long and enduring financial benefits. Emerging market data indicate that commercial buildings with lower energy usage—and the means to track energy usage by tenant—are more desirable, easier to lease, and have higher market value. A 2008 study examined Class A office leasing activity in 46 markets across the United States and found that green-labeled buildings achieve significantly higher rents—estimated at 7.3 to 8.6 percent for ENERGY STAR® labeled buildings and 15.2 to 17.3 percent for LEED® Certified buildings. Estimated occupancy levels were also higher by approximately 10 to 11 percent for ENERGY STAR labeled buildings and 16 to 18 percent for LEED-certified buildings.[5] However, submeters do involve an upfront cost of installation, integration, and training. Thus, it is important that the benefits of submeters over time exceed these upfront costs. Cost and benefit considerations are discussed further in Section 3 following the examination of four types of case studies.

2.1. Enabling Monitoring-Based Commissioning

A key benefit of submetering, albeit difficult to directly quantify, is the ability to enhance the performance of buildings. The availability of submetering data provided at more discrete levels can be used by automation systems as part of fault detection, thereby identifying problems with installed equipment before occupants observe adverse effects in utility bills. Additionally, submetering data may identify abnormal energy and water consumption as a result of system faults that may not be recognizable in the entire building's utility usage data. This detailed information enables more condition-based preventive maintenance in buildings, avoiding the higher costs typically incurred with deferred or unplanned maintenance. As part of a resource management plan, submeters can also provide the necessary data to document performance levels required by various energy labeling programs.

The case study presented below is based on in-depth benchmarking of a portfolio of 24 buildings located throughout the University of California and California State University systems. One specific California university building provides a good example of how submeters can provide critical operational insight.

***Case Study*[6]**

In this building, electric and gas meters were installed for a number of years, and they were manually read each month. After the existing electric and gas meters in the building were tied into the campus energy management system (EMS), and their energy use was trended, high nighttime electricity and natural gas use became immediately obvious. Much of the lighting was found to operate after hours and the air handlers operated continuously, even though the building was unoccupied at night. The chiller and boiler also operated at night, performing simultaneous heating and cooling; the manual usage readings had not triggered any alarms that would have revealed this problem. Having found these issues, the nighttime operation was easily corrected by reprogramming the EMS. The project also included installation of a BTU (British thermal unit) meter on the hot water output of the building's boiler. Initial readings revealed that the calibration factor used for the gas meter was not properly corrected for gas pressure, thus all of the historical gas meter readings were incorrect. The new gas readings are based on the correct multiplier and now compare properly with the metered hot water use.

2.2. Identifying and Monitoring Efficiency Retrofits

As part of the requirements outlined in EPAct 2005, EISA 2007, and EO 13514, Federal agencies must quantify their resource use and conduct comprehensive evaluations of potential energy and water retrofits. Often O&M managers rely upon energy and water audits to establish a snapshot of current resource se, peaks in resource demand, and, for electricity, the quality characteristics of the power delivered and consumed. Energy and water audits inform retrofit analysis and implementation by ranking which retrofit projects are most effective.

Energy audits are generally the first step a facility manager takes to determine the best investments to achieve greater building resource efficiency. Audit data provide a brief, short-term assessment of the performance of major systems and equipment. In order to rank the cost effectiveness of potential retrofit options, assumptions and approximations must be used to generate a rough estimate of energy or water savings over time. Often when using these retrofit estimates, they must be weighed against other operational realities, priorities, and ambiguities. These estimates can be significantly more accurate and realistic using submetering data on building and system performance over a longer time period, as it provides temporal, spatial, and seasonal usage measurements under actual operational and environmental conditions.

Similarly, water audits analyze a facility's water use, detect water losses, and identify ways to make it more water efficient. Audits often analyze domestic, sanitary, landscaping, and process water use and identify ways to increase a facility's water efficiency. Some local water utilities provide water audit services for no charge. The American Water Works Association (AWWA) provides free water audit software.[7]

Accurate measurement and verification is critical to ensuring the success of energy and water retrofit projects, particularly those undertaken using an energy service performance contract (ESPC) or utility energy service contract (UESC) with an energy services company (ESCO). The measurement capability of submetering verifies that the products and services offered in the contract are properly installed and operating as designed, prior to final acceptance. Additionally, under the ESPC, accurate measurement guarantees that the appropriate savings from the investment goes to the customer.

Energy Service Performance Contract (ESPC) / Utility Energy Service Contract (UESC)

An ESPC or UESC establishes a partnership between an energy services company (ESCO) and its customer for financing and implementing cost-saving energy-related improvements or renewable energy capacity, avoiding what can be a lengthy and time-consuming process to secure upfront capital funding. The ESCO pays the upfront cost of purchasing and installing new equipment, and the customer repays the ESCO over the life of the contract using cost savings resulting from the project. For an ESPC, the contract often specifies that customer retains savings in excess of the projected savings amount and the ESPC pays the customer the shortfall to that savings amount. Measurement and verification of system performance provided by submetering is central to both ESPCs and UESCs.

The use of ESPCs and UESCs can overcome a major barrier to submetering—financing the investment. For facility managers, securing funds for large capital improvements can be a lengthy, time-consuming, and difficult process. Due to this barrier, many Federal agencies are broadly embracing the use of ESPCs and UESCs, with over $1 billion in contracts signed in fiscal year 2010.

***Case Study*[8]**

Adobe Systems Inc. installed real-time main electric meters (revenue graded and Ethernet ready) in three high-rise office buildings, totaling just under 1 million square feet. The facilities included a cafeteria, fitness center, and 30,000 square foot data center. The meters were purchased and installed at a cost of $19,969. Adobe spent a total of $1.1 million on energy and energy-related projects, received $350,000 in rebates for these projects, and saved $1 million per year in reduced costs. Adobe earned a 122 percent return on investment, with a 10-month average payback. Shortly thereafter, Adobe installed real-time digital water meters (pulse meters) on cooling towers to record water loss through evaporation and water leaks in order to reduce its sewage treatment bill. The installation cost was $43,000, and the savings attributed to reduced sewer treatment charges resulting from water leakage totaled $12,000 that year.

***Case Study*[9]**

Southern Connecticut State University (SCSU) responded to a legislative mandate to accurately distribute costs associated with various university buildings in accordance with their source of funding. This initial project consisted of installing 14 submeters at various locations. The intent was to identify where older, obsolete equipment needed to be replaced. New high-performance equipment was installed along with digital controls to maximize energy savings. Rebates were applied for and granted for these projects; in some cases, the rebate paid up to 87 percent of the equipment premium. The average cost to purchase and install a submeter connected to the Direct Digital Control system was approximately $3,000.

Project	Initial Operating Performance	New Operating Performance	Reduction in Energy	Initial Upfront Costs and Rebates	Payback
Replace 300 chillers in science building	0.80 Kwh/ton	0.51 Kwh/ton	36%	Cost: $32,000 Rebate: $26,500	0.3 years
Replace 600-ton cooling tower	2–15 Hp single- speed drive fans	2–5 hp VFD drive fans	70%	Cost: $12,000 Rebate: $6,000	2 years
Install new 600-ton chiller complex	Initial design, 0.60 Kwh per ton, part load performance 0.59 Kwh per ton	0.51 Kwh per ton full load, 0.34 Kwh per ton part load performance with VFD drives on chiller	30%	Cost: $98,000 Rebate: $85,000 Calculated annual electrical savings: $22,500 (2002 Dollars)	0.6 years

***Case Study*[10]**

In 1977, the city of Boston, Massachusetts, could not account for the use of 50 percent of the water used in its municipal water system. After installing meters, the city undertook a vigorous leak detection program. Water that was unaccounted for dropped to 36 percent after metering and leak-detection programs were started.

2.3. Aligning Incentives and Enabling Behavioral Conservation

In addition to energy reductions from more efficient equipment operation or from replacing old equipment, it may be possible to reduce the energy use of a building by changing energy-use behavior of the building's occupants. Recent research in the residential sector, summarized in Appendix B, has shown moderate success with certain strategies aimed at changing the energy-related behaviors of residents. Literature reviews from the last ten years[11] find that energy reductions from a variety of feedback mechanisms can range from 0 to 20 percent. Reductions in energy use are achievable by providing occupants with appropriate forms of feedback on their energy use and highlighting ways to help them reduce their consumption. Research has also found that better forms of feedback are immediate and frequent, more disaggregated to different energy end uses, and comparative with both past energy use and with other similar users. However, there are challenges in applying these conclusions from the residential sector to commercial buildings due to obvious motivational differences among commercial occupants, who may lack direct financial incentives to reduce energy use. See Appendix B for more details.

Submetering can help encourage behavioral-based energy conservation in several ways. One is through aligning financial incentives directly to the occupants in multi-tenant buildings. By introducing building submetering of energy, building managers can bill each tenant based on actual energy and water use rather than on an average share of the total resource use. This provides tenants with an economic motivation to reduce their energy use, as they see an immediate financial gain from doing so. A similar strategy uses submetered data to provide rebates or prizes through competitions to various tenants or departments for decreasing resource use.

A major difference between the residential and commercial sectors is the level of system control of the average building occupant. Because commercial building occupants (including multi-tenant) generally do not control major energy subsystems, like HVAC and automated lighting, the potential reductions from a behavioral campaign are lower than in the residential sector. One exception is water systems, which may have more potential for behavioral changes, as occupants can alter or lower water consumption from improved sink and toilet flushing habits, (e.g., using less faucet water and one-flush per use). However, one important area of high impact is that of electrical plug loads, the fastest-growing source of energy consumption in all buildings. Occupant behavior such as turning off computers and other ancillary devices could potentially result from feedback mechanisms and other means using data from end-use meters.

***Case Study*[12]**

For multi-family residential dwellings, the State of New York and its energy research arm, NYSERDA (New York State Energy Research and Development Authority), have been prolific in facilitating the implementation of submetering and accurately documenting energy-saving results. In late 2009, NYSERDA released an updated survey of their submetering case studies for multi-family residential buildings. This report[13] confirmed their estimated savings of 18–26 percent through submetering using new and prior case studies on rent-stabilized apartments.

***Case Study*[14]**

About half of the 52-story Bank of America Building in San Francisco was submetered after energy managers learned that tenants exceeded by as much as three-fold their energy allowance of 3 kilowatts per square foot (kW/sq ft). More than 120 submeters were installed, and the property owner saved $1 million in excess energy usage in the first year alone. The cost of the submetering hardware and software in this application resulted in a payback period of days, not years, complemented by energy usage and cost savings of 30 percent per year.

***Case Study*[15]**

In July 2009, Vornado Realty Trust introduced submetering using hourly consumption data to a major tenant in New York City that occupies nearly 400,000 square feet of commercial office space. The tenant's average stabilized energy use was 1,080,600 kilowatt-hours (kWh) quarterly (4,322,400 kWh annually). Using the new hourly consumption profiles, the tenant identified energy-saving opportunities.

Based on a comparison of quarterly data comparing usage from April 26–July 26, 2009, to April 26– July 26, 2010, the tenant has saved on an annualized basis:

- 795,850 kWh in overall consumption, or 18 percent below the baseline
- 57 kW in reduced demand, or 8 percent
- An overall total savings of $160,000
- 266 metric tons of carbon equivalent (expressed as tenant carbon footprint)

2.4. Peak Demand Reductions

One potential ancillary benefit of submetering, at least for energy use, is the potential to decrease charges during peak electricity demand. Utility bills are calculated through both the overall use of energy and peak power demand. Thus, if utilities allow several billing meters on a campus to be aggregated and billed instead as a single "virtual" billing meter (a process known as "conjunctive metering"), thereby lowering the aggregated peak demand, energy bills can be lowered simply by having lower peak demand at any given moment.[16] This strategy of reducing peak demand through billing at larger spatial levels is known as "virtual aggregation." Other potential benefits of virtual aggregation include reduced customer service charges and lower rate structures for larger loads.

Case Study[17]

At a large research university, the most common use of submetering data was to bill separate departments for their energy usage. As the metering program matured, the departments received more accurate billing data on energy consumption and demand. The improved accuracy resulted in increased charges to certain departments, which led to their reevaluation of energy use and improved conservation so to lower their utility bills. To achieve further savings, the university leveraged the metering system by negotiating an agreement with the power utility, allowing the utility's meters to be used as submeters as part of a "virtual aggregation" program to save on peak demand charges.

The virtual aggregation of submeters installed at the building level reduced the cost of demand (kW) by at least 10 percent in all 30 buildings involved. These savings were passed on to the university departments. Submeters were also used to more accurately estimate energy and water loads, thereby reducing distribution system design requirements and lowering construction costs for new buildings.

2.5. Combination of Benefits from Submetering

Many of the benefits and uses of submetering data can be combined to achieve even more savings. As reported in the previous case study, owners may introduce submetering for one purpose, such as billing, and effectively use the data for another purpose, such as design and retrofit analysis, motivating behavioral change, or optimizing system operations. This combination of benefits is well illustrated by a second example from the same report.

Case Study[18]

The research university began submetering its campus more than 15 years ago, driven by pressures to reduce overall expenses and to allocate indirect costs to the appropriate research activities. Through an energy awareness program, the university offered departments a chance to receive payments of up to 30 percent of the savings achieved, if they cut costs relative to baseline energy usage.

- After 18 months of billing the departments for their usage, the school had reduced its usage from about 44 million kWh to 40 million kWh, saving about $300,000 per year. These energy and cost reductions were accomplished through improved operations and maintenance procedures that eliminated wasteful practices in energy intensive research departments.
- The study identified several energy retrofit projects costing about $2.5 million, with a four-year simple payback, resulting in a second round of major energy cost reductions totaling about $625,000 per year.
- The study evaluated a cogeneration capability, pressuring the utility and the utility commission to agree to secure lower, long-term rates for the school.
- From all phases, submetering has saved the university about $1 million per year for the past 10 years.

3. Economic Considerations

3.1. Life Cycle Cost Analysis

All energy-related investments, such as submetering, should be evaluated and prioritized using life cycle cost analysis (LCCA). The purpose of LCCA is to ensure that the investor procures building improvements that represent the best value over the useful life of the asset rather than simply those with the lowest upfront costs.

Following EISA 2007, managers of Federal buildings are now required to calculate life cycle costs[19] on a 40-year basis using the sum of the present values of:

- Investment costs, less salvage values at the end of the assets'expected useful life
- Non-fuel operation and maintenance costs
- Replacement costs less salvage costs of replaced building systems
- Energy and/or water costs

An investment is "life cycle cost effective" when the net present value of the input variables produces net savings. Conventionally, LCCA is expressed using one of three standard measures:

- Savings-to-Investment ratio (SIR)[20]—An SIR estimate greater than one indicates an economically worthwhile project.
- Present Value Net Savings (PVNS)[21]—Net project savings, discounted over time, greater than zero indicates economic value.
- Adjusted Internal Rate of Return (AIRR)[22]—AIRR is the annual yield from a project over the study period that takes into account its costs and benefits, the time value of money, and reinvestment of interim receipts. If the AIRR is greater than the minimum attractive rate of return, the project is economically worthwhile.

All three of these measures, when computed properly, will yield similar outcomes when applied to the same project; in practice, however, these standardized systems do not always support effective whole-building management.

For a number of reasons, many benefits of building system retrofits are difficult to accurately quantify. In some cases, a single management action to reduce energy consumption in a given building system will impact other systems. For example, a submetering report that indicates excessive use of lighting during periods the building is unoccupied can lead to a reduced lighting load, which may in turn affect the heating and/or air conditioning load. In addition, actions that improve building operations can enhance the comfort of building occupants, producing indirect benefits (both monetary and human) such as improved satisfaction and productivity. Similarly, other non-monetary benefits, such as reduced power plant emissions, generally do not factor into LCCA. Such benefits, while difficult or impossible to effectively integrate into current LCCA methodologies, represent important aspects of submetering's total value proposition and merit inclusion in the decision-making process.

3.2. Distribution of Costs and Benefits

The distribution of the costs and benefits of submetering represent another barrier, often referred to as "split incentives." For example, in buildings where owners sublet space to tenants, the owners often are responsible for installation of submeters and related software and equipment, but tenant behaviors ultimately affect changes in building resource use. The costs to install the submetering systems that lead to these changes fall to the owner, while the energy cost savings primarily benefit the tenants through reduced utility bills.

Generally, the cost-benefit calculus of submetering is similar to many building retrofits, with high upfront costs offset by long-term operational and cost benefits. The costs and capabilities of meter systems can vary widely and may be subsidized by Federal, state, local, or utility incentive programs (see Appendix C). Meters should be installed and operated by trained and qualified technicians, generating both an additional first cost as well as an ongoing maintenance cost. In addition, for metering to be useful, data must be collected from each device and analyzed to provide insight into the operation of the building—associated software, devices, displays, and networks for data collection and analysis are another first cost. Finally, if the data culled from submetering are to provide any benefit, it is most often because other actions have been taken to modify the building's energy or water usage, and a wide range of options are available as described in this paper.

Despite these challenges, building owners can reap potential benefits from submetering in several ways. First, owners of submetered buildings may negotiate better rates with utility companies for the resource(s) used, benefiting both owners and tenants, and may be combined with other strategies such as demand-side peak reduction. Owners of buildings with tenant-level electrical submetering can serve as the local power distributor, buying energy at a wholesale rate and billing tenants based on actual submetered usage that includes a reasonable rate of return. In some jurisdictions, complex rules govern how building owners are to distribute the actual cost of energy use to individual tenants. Adequate incentives must exist for building owners to make the necessary capital improvements to their buildings while at the same time protecting tenants' interests.

Second, without effective conservation measures to quell tenants' uncontrolled energy use, building owners may be required to invest in expensive capacity-extending equipment, such as larger electrical riser systems, that distribute electricity throughout the building. In this scenario, the submetering investment may yield savings through reduced or avoided capital outlays. Similarly, a more refined understanding of system performance using submetering data may result in changes in building operations through the use of intelligent controls that automatically adjust to environmental requirements. These types of strategies can save building owners money and time while also improving indoor environmental quality for tenants.

Finally, the use of secondary strategies and approaches may encourage building operators and occupants to adopt more efficient O&M strategies that reduce their resource consumption and improve efficiency. This is done by the building's submetering operator providing accurate and specific information on resource use to the building's O&M technicians and occupants, identifying high consumption areas/ systems, encouraging conservation on a regular basis, and incentivizing reduced resource consumption, such as lower utility bills, to both groups.

Submetering at the tenant level, though, may not benefit all tenants equally. In buildings that were previously metered by a single master meter with costs shared proportionately, tenants with higher resource use will face higher utility bills.

Case Study[23]

In the NYSERDA report on multi-family dwellings, ratios of the highest user of electricity in a typical apartment complex to the average user of electricity in the same apartment complex typically calculated to a ratio of approximately 3:1. Analysis of usage confirmed that approximately 10 percent of the building's apartments consumed typically 20 to 25 percent of total apartment usage, and that approximately 70 to 80 percent used less than half of the total apartment usage, a finding that has been referenced in prior NYSERDA-funded studies.

While the case studies presented thus far are not intended to imply that submetering guarantees positive returns, they do suggest that a careful analysis of costs and benefits show that submetering may be economically worthwhile. At the same time, they reveal the need for well-documented, economically sound, comparable studies based on standard measures of worth (see Section 5, Next Steps). Only then can the economic discussion progress from anecdotal case studies toward the rigorous analysis necessary to inform future submetering user communities.

Operationalizing LCCA in the context of institutional operations and resource limitations creates other challenges. Although Office of Management and Budget (OMB) Circular A-11 (Preparation, Submission, and Execution of the Budget) requires life cycle cost reporting for Federal agencies, practice managers tend to do so mostly for information technology investments, not buildings, for a few reasons. First, agencies often separate capital budgeting from operational budgeting, limiting the ability to effectively link first costs to future operational costs. Moreover, agencies are often undercapitalized for high-cost, high-impact projects because the project budget estimates often exceed their limited funds. In some cases, managers can more easily justify investments based on upfront cost data rather than on subjective environmental, operational, or social variables that are difficult to quantify or—in the case of future costs and savings—difficult to forecast.

4. Submetering Technical Details

For electrical systems, meters can be installed to track whole-building energy use (e.g., utility-owned advanced meters), sub-panel energy use (e.g., a lighting or process circuit), or a specific end use (e.g., an occupant's space, a motor, or a chiller). For water, natural gas, and other flow-related applications, meters are typically in-line installations on distribution lines using positive displacement, insertion turbine, or pressure-related techniques. Depending on the need, any of these meters will vary in size, type, output configuration, accuracy, and price. A definitive source for additional details beyond those provided below can be found the U.S. Department of Energy (DOE) Federal Energy Management Program's *Metering Best Practices Guide*[24] and *Operations and Maintenance Best Practices*.[25]

The four predominant levels of resource metering[26] are:

- One-time/spot measurement
- Run-time measurement
- Short-term monitoring
- Long-term monitoring

One-time or spot measurement applications are useful in many "baseline" activities to understand instantaneous energy use, equipment performance, or loading. In addition, they can measure changes in consumption following energy efficiency upgrades. Run-time measurements are often used to establish hours of operation for devices or systems as part of reducing energy consumption through shortened operation times. Short-term monitoring combines both duration of operation with energy consumption to establish a time-series record of energy or resources used. Long-term monitoring also makes use of duration of operation and energy consumption, though the equipment used is often installed on a permanent basis, and measurements are taken at a prescribed frequency and resolution to meet the established energy monitoring requirements. This last approach is the focus of the "Energy Use Measurement and Accountability" portion of EPAct 2005. For more information on the fundamentals of submetering, see Section 1.2.

The process for metering and submetering consists primarily of:

- Determining what should be measured in any particular building or location.
- Measuring the physical properties of flow through the distribution system to determine how much of each resource is being consumed.
- Collecting data at predetermined intervals and storing those measurements.
- Analyzing the measurements to determine how much of each resource is being used.
- Making informed decisions based on all information provided.

4.1. Meter/Submeter Performance Metrics and Attributes

Performance measures for meters and submeters include the metrics listed below as identified in the *Metering Best Practices Guide*.

> **Accuracy**—Accuracy is the difference between a measured value and the true value. Published accuracies often will, and should, be referenced to specific calibration procedures, including equipment traceability to National Institute of Standards and Technology[27] equipment and procedures.
>
> **Precision/repeatability**—The precision or repeatability of a measurement entails the ability to reproduce the same value (e.g., temperature, power, flow rate) with multiple measurements of the same parameter, under the same conditions.

Turndown ratio—The turndown ratio refers to the flow rates over which a meter will maintain a certain accuracy and repeatability. The larger the turndown ratio, the greater the range over which the meter can measure the parameter.

Beyond performance metrics, other criteria useful for meter selection include:

Ease of installation—Regardless of technology, meters come in many different sizes and shapes, need a variety of inputs, and offer a variety of outputs. For specific make-and-model decisions, it is important to understand any size and weight constraints, needs for specific diameters (or lengths) of straight pipe upstream and downstream of the meter, specific electrical and communications needs, and the overall environment the meter will operate in.

Ongoing operations and maintenance—The lowest-price metering technology may not be the best choice if it has high associated maintenance costs (e.g., frequent service, recalibration, sensor replacement). As with most capital purchases, a life cycle cost approach (including all capital and recurring costs) is recommended for decision making.

Installation versus capital cost—In some situations, the cost to install a meter can be greater than the capital cost; this can be true when system shutdowns are necessary for meter installations, or where significant redesign efforts are needed to accommodate a meter's physical size, weight, or required connection. In these cases, decision makers should consider alternative technologies that may have a high first cost but a much lower installed cost. A good example is the use of non-intrusive metering technologies (e.g., ultrasonic flow meters) that typically have a high capital cost but often a significantly reduced installed cost.

Energy and water meters all require periodic maintenance. See the *Metering Best Practices Guide* for a complete description of periodic maintenance requirements.

4.1.1. Electrical Submetering

Electrical meters found on many homes and businesses are one of three types: *mechanical meters,* where electricity use is measured by the movement of a mechanical dial; *electromechanical meters,* which are similar to mechanical meters but transmit data by electronic or pulse output; or solid-state/ digital meters—also called *advanced meters*—that use non-mechanical sensors to provide data in a variety of digital formats.

Advanced electrical meters represent the highest-level components of a submetering strategy and satisfy the "Energy Use Measurement and Accountability" portion of EPAct 2005 for Federal facilities. Automated meter reading (AMR) systems, both wired and wireless, are heavily used due to Federal mandates, lower life cycle cost, and greater capacity for data capture and transmission. They contain integrated circuits with current and voltage transformers, on-board memory, and communication technology. They also have the capability to measure and record interval data (at least hourly for electricity) and to communicate it to a data collection system. Advanced meters range from the familiar circular socket-based style to an array of rack- and panel-mount configurations. Their accuracy ranges from 0.05 to 3 percent at a cost of $400–$3,000.

The three main types of electrical devices for electricity submetering (see Table 1) are electromechanical/ solid-state feed-through and current transformer (CT) meters, and the electronic non-socket meters. Feed-through and CT submeters are both socket-type meters.

Feed-through Meter

In the feed-through installation, the meter is physically connected between the receptacle and the end load; the end load plugs into the meter, which in turn plugs into the receptacle. The main advantages of feed-through meters are their accuracy and precision, both increasingly critical because of the need to measure low power or standby power levels and the richness of data provided, which can include voltage, current, frequency, power, energy, and power factor. The main disadvantages are the additional utility room space requirements and the need to temporarily unplug the load to install the meter. This configuration is generally acceptable for the vast multitude of plug loads; however, it may be a problem for loads that require a planned turn-off or shut-down with proper backup (e.g., computing or telecommunications equipment, or medical equipment).

Current-Transformer Meter

CT submeters are used with loads in excess of 400 amps—too high to directly apply to measuring instruments. A current transformer produces a reduced current accurately proportional to the current in the circuit, although installation is more complicated because they must be mounted in a large cabinet for electrical code compliance. Both feed-through and CT meters occupy between four to ten square feet and must be installed in a utility room.

Electronic Non-Socket Meter

The main advantages of electronic clamp-on CT probe meters are their availability and mature technology, as well as their mature data collection infrastructure. Clamp-on CT probe meter installations are less popular for power data collection due to the need to properly "split" the power cord for clamping the CT to a conductor. Other disadvantages are that these meters are generally better suited for higher loads, like those available as an aggregate at the panel or subpanel level, and they are limited to AC loads only. New device features appearing on the market continually improve their application features.

All three devices provide a range of measurement and reporting capabilities. Hardwired sensors utilizing a dedicated wire from the measuring device(s) to the data acquisition are probably most common, but information can also be transmitted via power line carriers or radio signals.

Determining the types of data to collect and store, and at what frequency, can be a challenge when developing a plan and budget for electrical submetering. The frequency at which the electrical data need to be collected depends mainly on the individual end load being measured, which often dictates the minimal resolution needed for properly analyzing energy and power consumption of the end use. The proper submetering configuration, data types and frequency, software, and analysis needed to achieve prescribed operational goals can be complicated and require in-depth examination of the buildings, systems, and desired outcomes.

As the number of possible submetering configurations and applications is quite large and dependent on the building design and use, it is impossible to provide adequate details for the building designer in this paper. Please see "Submetering—A Practical Approach" by Mane (2005) for simplified examples of submetering applications and hardware requirements organized by market segment, including commercial multi-tenant buildings, residential multi-tenant dwellings, commercial retail, industrial/ manufacturing, and institutional buildings.[28] See also Appendix C for examples of submetering configurations and device pricing.

4.1.2. Electrical Submetering of Plug Loads

As mentioned earlier, electrical meters can be installed to track whole-building energy use (e.g., utility meters), sub-panel energy use (e.g., a lighting or process load), or a specific end use. In the latter, the end-use electric load is sometimes referred to as a plug load or miscellaneous electrical load.

Table 1. Types of Electrical Submeters—Example Comparison

Specifications	Submeter Type		
	Electromechanical/Solid State		Electronic Non-socket
	Feed-Through	Current Transformer	
INSTALLATION			
Installed Cost (estimated)			
-Stand alone, up to 320A, 3Ø	$1,000	N/A	$700
-Stand alone, over 320A, 3Ø	N/A	$2,000–$5,000	$800
8-Meter Unit, 200A, 3Ø	$16,000	N/A	$5,500
Installation Time	2–3 Hours	6–8 Hours	1 Hour
Power Interruption	2–3 Hours	6–8 Hours	None
Amperage Limitations	320 Amp, Max.	None	None
Space Requirements	2 Square Feet	11.7 Square Feet	0.25 Square Feet
Installation Location	Utility Room	Utility Room	Anywhere

Table 1. (Continued)

FEATURES			
Multiple Meter Units (MMU)	Yes	Yes	Yes
Size of 8-Unit Cabinet	18.1 Square Feet	18.1 Square Feet	2 Square Feet
Digital Readouts	Optional/Yes	Optional/Yes	Standard
Reset Capabilities	No/Yes	No/Yes	Standard
Multiple Load Monitoring	No	No	Yes
Subtractive Load Monitoring	No	No	Yes
Monitor Specific In-Panel Circuits	No	No	Yes
Amperage Modification in Field	No	w/ CT Change	Yes
Meter UL listed	No	No	Yes

ENHANCEMENTS			
Digital-to-Analog Profiles	Yes	Yes	Yes
Pulse Outputs	Yes	Yes	Yes
Timed Metering	Yes	Yes	Yes
Software Monitoring	Yes	Yes	Yes
Upgradeable in the Field	No	No	Yes
Power Quality Functions	Available	Available	Yes
Net-Metering Capability Form C	Yes	Yes	Yes
Control Relay Output	No	No	Yes

Source: E-Mon, LLC.

Plug loads are generally defined as all non-main building electric loads; that is, all electric loads except those related to the main systems for heating, ventilation, cooling, water heating, and lighting. Plug loads account for an increasingly large portion of commercial and residential electricity consumption.[29] This trend is mainly due to an increase in both the number of types of end loads and the number of loads in buildings. Typically, plug loads fall into a lower power consumption category, and therefore clamp-on CT probe meters are still prevalent for subpanel metering and not plug load metering.

Plug load metering is also used to disaggregate data obtained through subpanel metering. For example, a single, submetered breaker may serve both lighting and plug loads, when only metering for lighting is desired. Plug load meters installed on individual end uses served by the same breaker may be used in conjunction with subpanel submetering to disaggregate power measurements by subtracting the plug load from the breaker's metered data—thus yielding a more accurate lighting-only measure.

In general, however, plug loads require a relatively high frequency of data collection when compared with other submetering activities. When plug loads are pervasively submetered with multiple parameters being measured (e.g., voltage, current, power, frequency, power factor), the energy and power analysis requires a rigorous, reliable, and scalable data-collection infrastructure able to handle the significant data flow. Pervasive plug load metering is currently pushing the available technology and data collection infrastructure to its limits. In addition, because of the high installation costs associated with the pervasive metering, wireless meter configurations are often favored, adding to the complexity and challenges of the submetering solution. As a result, deployment of large-scale submetering of plug loads is not currently widespread.

4.1.3. Potable Water Submetering

Throughout a building or campus, water is distributed by a series of pressurized distribution pipes or lines. A vast majority of buildings today measure water once upon entering the site, and do not measure successive distribution points or end uses. There is no real equivalent of an "electrical panel" for water distribution in most buildings, making meter placement more problematic. The technology and applications supporting water metering are therefore significantly less mature or pervasive than those for energy, though metering water and recording water use can improve resource conservation and enable system diagnostic capabilities.

The physical properties associated with potable water lend themselves to a wide variety of metering technologies. The lack of high temperature and/or two-phase flow allows the use of metering technologies not suitable for other applications (e.g., steam or high-temperature hot water). Because potable water metering is typically focused on the quantification of flow volume, and not energy content, lower-cost options can be used. The option selected will depend on a number of factors including, but not limited to, current design, budget, accuracy requirements, minimum flow, range of flow, and maximum flow. In general, volumetric water metering designs can be broken down into three general operating designs: positive displacement, differential pressure, and velocity including ultrasonic meters. The Metering Best Practices Guide provides more detailed information.

At the whole-building level, where nutating disk meters are the most common, calibrated pulses are the typical data output and are relatively easy to work with, though other output options are available (4–20 milliamp, 0–5 volt, Modbus, etc.).

Potable Water Meter Type	Application	Accuracy	Cost
Nutating Disk	Up to 3" connections	0.5%–1.0%	\$50–\$400
Orifice	0.25"–4.0"	0.25%–2.0%	\$1,000–\$5,000
Venturi	0.25"–4.0"	0.25%–2.0%	\$1,000–\$5,000
Turbine	2.0"–20.0"	0.5%–1.0%	\$500–\$5,000

Measuring the properties of water flow, such as flow rate and temperature, is relatively straightforward. The methods may be invasive (cutting the piping and installing a meter) or non-invasive (no cutting required). Some installations may require stable laminar flow and mandate a straight length of pipe 15 to 25 pipe diameters upstream and 5 to 15 pipe diameters downstream of the meter. Like electrical metering, the signals from these measurement devices can transmit to an appropriate data acquisition system in many ways. Hardwired sensors using a dedicated wire from the measuring device(s) to the data acquisition system are probably most common, but information can also be transmitted via power line carriers or radio signals. Using a flow totalizer system, meter data can be combined from different locations via a data network to provide functions of flow control and measurement through remote online management. Using these systems, flow measurement, pressure, and temperature are measured and can determine the density of the measured object and accumulate the flow from multiple locations over time.

Building owners or occupants interested in measuring water consumption will need to understand the entire water distribution system, choose where they want to measure, and then install appropriate meters at that location. Like electrical metering, deciding what data to collect and store, and at what frequency, can be a challenge. The signals from water measurement sensors are real-time, meaning that instantaneous values are available. The data acquisition system must be able to process and store these data and measurements to facilitate accessibility for analysis.

4.1.4. Natural Gas and Steam Submetering

Natural gas and steam submetering share similarities to water submetering. All three involve fluid flow through pipe distribution systems, though with natural gas and steam metering the "fluid" is in the gas phase and the changes in temperature and pressure can result in metering inaccuracies. Also, like water, natural gas and steam distribution systems involve life-safety issues, and gas flow consists primarily of the same three categories of measurement: positive displacement, differential pressure, and velocity.

At the whole-building level, where positive displacement diaphragm and rotary meters are the most common for natural gas, calibrated pulses are the typical data output.

Natural Gas Meter	Application	Accuracy	Cost
Diaphragm	0.75"–2.0"	1.0%	$150–$500
Rotary	1.5"–4.0"	0.25%–0.5%	$300–$500
Orifice	0.25"–4.0"	0.25%–2.0%	$1,000–$5,000
Venturi	0.25"–4.0"	0.25%–2.0%	$1,000–$5,000
Annubar	2.0"–100.0"	2.0%	$1,000–$3,000
Turbine	2.0"–20.0"	0.5%–1.0%	$500–$5,000
Vortex Shedding	1.0"–12.0"	1.0%–2.0%	$4,000–$6,000

Steam flow metering presents challenges in accuracy and maintenance. Most steam meters measure a volumetric flow due to the two-phase physical properties of steam, which may impair accurate measurement of flow rate. Steam use is less pervasive than electricity, though metering can provide valuable diagnostic capabilities in certain applications. Meter maintenance is also an issue depending on the meter technology and installation and the quality of the steam transferred.

Steam Meter Type	Application	Accuracy	Cost
Orifice	0.25"–4.0"	0.25%–2.0%	$1,000–$5,000
Annubar	2.0"–100.0"	2.0%	$1,000–$3,000
Turbine	2.0"–20.0"	0.5%–1.0%	$500–$5,000
Vortex Shedding	1.0"–12"	1.0%–2.0%	$4,000–$6,000

Measuring the properties of natural gas and steam is relatively straightforward but almost always invasive (requiring cutting of the piping and installation of a meter). Due to the flammable nature of natural gas and the high temperature of steam, meters for natural gas and steam should always be installed by qualified professionals. Like electrical and water metering, the signals from these measurement devices can transmit to an appropriate data acquisition in many ways. Hardwired sensors using a dedicated wire from the measuring device(s) to the data acquisition system are probably most common, but information can also be transmitted via power line carriers or radio signals.

Some installations may require stable laminar flow and mandate a straight length of pipe 15 to 25 pipe diameters upstream and 5 to 15 pipe diameters downstream of the meter. As with water metering, there may not be a natural location equivalent to an electrical panel to conduct natural gas or steam measurements. A building owner or occupant interested in measuring natural gas or steam consumption will need to understand the entire natural gas or steam distribution system, choose where they want to measure, and then install appropriate meters at the appropriate locations.

4.2. Installation and Integration of Submeters

Installing submeters in a building requires careful design analysis to meet a project's energy efficiency goals and objectives. As mentioned earlier, submetering by itself has relatively little impact on overall energy or water consumption—it is a tool to help building owners and occupants understand and control their usage. Proper submetering designs must include the specification of the level of submetering required, the types of data to be collected and used to reduce consumption, and the software, analysis, and communication tools necessary to manage energy usage by systems, in order to provide information to operators or occupants and/or to control the building's energy and water use directly. Submeters can interface with the building automation system, not only making it possible to integrate the control logic with the meter but also providing access to the building automation system tools for trend analysis, report generation, and user information display.

In the Southern Connecticut State University (SCSU) case study referenced earlier, submeters were installed on the secondary side of the switchgear in the building or facility. SCSU reported that besides having a lower cost, secondary submeters were much easier to install, and electricians could install them in about three hours while performing cleaning and maintenance of load centers concurrently or separately. The installation involved connecting three current sensors on the electrical feeds. Because the submeters were connected to a digital control system, the installation did not require a separate data network or other major interior changes in the building or electrical distribution system. The average cost to purchase and install a submeter connected to the data control system was approximately $3,000.

4.2.1. Selected Issues for Special Building Configurations

Special-use buildings may present other unique opportunities and challenges. Resource intensive facilities such as hospitals, healthcare facilities, and laboratories are highly complex buildings. These special building types offer even greater opportunities to improved performance. In the case study below on building commissioning for new and existing buildings, median payback periods were considerably shorter for energy intensive facilities than for other types of buildings due to the larger economic value per percentage improvement in building efficiency. Submetering is an integral component to monitor-based continuous commissioning approaches for the next generation of high-performance buildings.

Case Study[30]

In a study of 643 buildings, commissioning data revealed over 10,000 energy-related problems, resulting in 16 percent median *whole-building* energy savings in existing buildings and 13 percent in new construction, with payback time of 1.1 years and 4.2 years, respectively. Median payback periods were considerably lower for energy intensive facilities, including 0.5 years for laboratory spaces and 0.1 to 0.6 years for healthcare facilities.

Health care institutions are both energy- and water-intensive facilities—hospitals and healthcare facilities are consistently within the top 10 water users in their communities. Typical water use per capita in hospitals ranges from 40 gallons per day to 350 gallons per day.[31] Water use at a medical facility depends upon the services provided, in-patient vs. out-patient visits, staff attendance, equipment used, age of the facility, and periodic maintenance practices followed.

Case Study[32]

Boston's Norwood Hospital engaged in a major program of implementing water efficiency measures. By eliminating seal and cooling water on medical air compressors and vacuum pumps, 8.5 million gallons of water was saved. Using re-circulated seal and cooling water for four vacuum pumps and one medical compressor as well as removing a vacuum pump resulted in a net annual savings of 8.5 million gallons. Norwood also installed a number of sanitary retrofits—aerators and flush valves saving 3 million gallons per year. Replacing the flush valves on toilets and urinals, and installing low-consumption aerators on all lavatory faucets resulted in another savings of 3 million gallons per year. A water refrigeration system for the morgue was replaced with an air-cooled unit, saving 2.1 million gallons. Water use dropped from 51.2 million gallons in 1991 to 36.6 million gallons in 1994. Project Cost: $5,500, project savings: $13,750, payback period: 0.4 years.

In hospitals and healthcare facilities, many challenges to improved resource efficiency exist. Energy- and water-intensive medical equipment are used around the clock, thus requiring they remain "powered on" 24 hours a day. Energy and water consumption are driven by hours of operation, number of patients, staffing intensity, and healthcare functions; hospitals and tertiary care facilities generally have high consumption rates than long-term care and mental health facilities.

Often medical devices are not energy rated and are therefore absent of quantified energy performance requirements and data. Such devices represent *unregulated loads* to the building and currently are not covered by building efficiency codes. Further, some of these devices may be plug loads that are shifted from room to room as needed (e.g., blood pressure and intravenous flow control machines).

Installation procedures in a hospital environment are also more challenging. Concerns about infectious disease in an operating healthcare facility require complicated infection-control and scheduling procedures. Access to utility areas above the ceiling or in pipe chases is controlled to prevent potential contamination from accumulated substances and debris. Scheduling projects such as meter installation is often complicated by the necessity of continuity of operations, and the use of wireless communication devices must not affect sensitive medical equipment that can be compromised by radio frequencies.

Computer and data communications in healthcare facilities require high levels of security protection due to privacy requirements related to sensitive patient information. Mechanical and electrical systems in facilities with special-use areas can be much more complex and diverse, with strict requirements on air pressure relationships, air exchange rates, and temperature/humidity control. Facilities also have specialized emergency electrical systems, more extensive fire and safety requirements, and other specialized building requirements (e.g., dedicated air conditioning systems). These varied requirements complicate the ability to implement submeters in a manner that will provide meaningful data. In these types of facilities, the cost-benefit analysis must include life-critical requirements, functional needs, and standard operating procedures for healthcare facilities.

Other special building types include laboratories, specialized research facilities (such as particle accelerators), and energy-intensive areas such as data centers and clean rooms. In the report cited in the previous case study, energy savings in a particle accelerator facility, attributed to a more efficient operation of chillers, resulted in a 46 percent reduction in energy and an avoided capital cost of $120,000. This report presents other dramatic case studies on the building's monitor-based commissioning using submetering data. As an interesting aside, while these special-use buildings strictly control their working environments, regularly scheduled O&M procedures frequently do not address system energy efficiency and performance requirements.

Key issues in implementing submetering in existing buildings are the inconvenience and risks presented to building occupants. The extra meters required for submetering may require rewiring, installing data networks, and other intrusive measures, which could be especially problematic in historic buildings or buildings in which meter installation may present occupational hazards, such as exposure to asbestos. To partially mitigate these problems, wireless communication technology and data transmission can be used to in place of a new data network for the building. Current wireless communication networks though may, if not properly configured, introduce network security issues related to data integrity and information security when connected to standard data networks. See also Section 4.3.2.

One of the biggest challenges with submetering lies with *in situ* electrical distribution systems, where for a specific end use the electrical feeds may be dispersed throughout multiple electrical panels. For example, to know how much power hallway lights use in a building, where the lights are connected to 15 different breakers in 10 different panels on three different floors, it is difficult and costly to measure the power through a single submetering system.

As a result, this issue is driving changes to building electrical codes that would require end uses to be grouped at the electrical panel. Eventually, as submetering for natural gas and water become more common, the issue is likely to arise there as well.

Older multi-tenant buildings that previously used a single meter may have electrical layouts that combine loads from multiple areas, further complicating the process of installing measurement devices dedicated to individual areas. However, eliminating "cross-wired" spaces is essential to accurately and fairly bill tenants for their energy consumption. To help building owners learn more about submetering, the Utility Management & Conservation Association (formerly known as the National Submetering and Utility Allocation Association) offers helpful resources and information.

4.2.2. Integrating Submetering with Existing Operations

Submetering is just one important component to the overall operation and maintenance of a building. A submetering solution must be carefully defined during the specification and contracting phases to include adequate integration with existing building management systems and O&M procedures, selecting appropriate equipment and software systems, ensuring adequate training, testing, and commissioning of the installation, and providing appropriate maintenance and service costs and schedules. Ideally, the submetering solution should be designed as part of an ongoing program of continuous monitor-based commissioning. A building's O&M staff and occupants are integral to the submetering solution, as human behaviors often drive building energy use and impact overall success. Submetering may also be part of a larger effort to document and report on resource use or on evaluating building design and retrofit alternatives.

Currently, there exist a number of challenges in implementing advanced metering and submetering solutions in new and existing buildings. The first challenge concerns properly defining Requests for Proposals (RFPs) that rigorously list the performance specifications of the equipment, installation requirements, testing and commissioning, data requirements and protocols, and interfaces to new and existing building/energy management systems (BMS/EMS). To facilitate correct operation, EMS systems should be integrated into the building's automation or continuous recommissioning supervisory systems to detect failure of individual devices or meters. Ownership of the building's advanced meters can also present challenges; for advanced meters owned by utilities or other entities, agreements should be established to outline the procedures for changing out meters or altering operating specifications.

Human factors play an important role in successfully installing and operating a submetering system. Once meters are installed, trained O&M technical personnel, both on staff or under contract, must maintain the devices and ensure they operate with high levels of accuracy and reliability. Dedicated staff is essential to operate the EMS system and to make any software and programming changes, so the building management system can efficiently respond to setpoints and alarms. To meet these needs, appropriate resources must be dedicated to train staff or to contract these services, and budgets must account for replacing and repairing submetering components as dictated by maintenance schedules. Incentives to O&M staff may be required to ensure that the submetering devices remain fully operational and that they are used as designed. Additional incentives may be required for occupants to change their behaviors and effect conservation and resource savings.

Submetering systems must be designed, implemented, and maintained as production-quality building management systems, as opposed to experimental tools. High-levels of reliability, information assurance and security, established data acquisition/ archival procedures, and periodic system maintenance requirements are essential to acquiring and using the submetering data over the long term. Operational responsibility for submetering systems must be clearly designated to ensure availability, accessibility, and dependability, as occupant behaviors may obscure results, transfer loads to unmonitored locations, or result in disproportionate use of other resources. Other rules and regulations may also conflict with the implementation of submetering technologies (e.g., bans on wireless communications).

4.2.3. Integration with Whole-Building Automation Systems

One relatively easy way to collect submetered data is to connect the submeters to an existing building automation system. Automation systems collect and process system information on the performance of a building and, with appropriate connections and programming, they can compile and use submetered data as well. Since most measurement technologies for electricity, water, natural gas, or steam can be set up to provide output in the form of electrical signals, those signals can be fed into a whole-building automation system for collection and processing. Likewise, it is also possible to take specific control actions or to generate event reports based on meter data and signals.

Standards for interconnecting components in building automation systems have been developed that ease the integration of submeters. For example, Open controls protocols BACnet[33] and LonTalk,[34, 35] two standardized data communication protocols for building automation and control networks, have specific features designed for interconnection with meters. The meters can directly support the communication protocol, or pulse outputs from meters can be used by the building automation system to collect data. Most submeters can be readily integrated into building automation systems through one of these methods. Standard data protocols facilitate the identification of data from different meters as well as its consolidation into analysis and monitoring software. Once a meter is integrated with the building automation system, all of the trending, analysis, and user display tools of the system become available for use.

For many organizations, when a new data network is connected to an existing, secured data network, it must possess appropriate information assurance and security characteristics. Standard security standards, such as IPsec and Kerberos, were designed to operate only on TCP/IP networks and as such do not contain the necessary features to support control networks and protocols. To achieve these network security requirements, the control protocols have been extended with a set of network layer security messages that provide data confidentiality and integrity, device authentication, data hiding, and user authentication. See also Section 4.3.2.

Whole-building automation systems are by no means ubiquitous, and existing buildings may not have systems in place that can integrate submetering data. Installing and operating new submetering systems can be complicated and should incorporate the utility's ownership and maintenance of the primary meter, as the utility's O&M activities may pose a risk to the reliability of the submetering and data collection system. Furthermore, conflicts may arise between those responsible for the submetering system and coordination with multiple entities (occupants, data network managers, building owners, utilities) that can affect optimum system operation.

4.3. Tools for Analyzing Submeter Data

The measurement capability provided by submetering provides building owners with a structured method to reduce resource consumption and to systematically identify and eliminate operational variations that cause resource waste. In order to determine the improved efficiency of a building or of a particular retrofit, the building's performance data must be compared relative to its prior performance and to other buildings of similar design and use.

4.3.1. Benchmarking and Baselining

Energy and water use benchmarking is a process that compares the energy and water use of a building or group of buildings with other similar structures or looks at how energy and water use varies from a baseline.[36] Benchmarking energy performance helps identify best practices that can be replicated, either within a building or across a portfolio of buildings, using performance data from similar buildings. Benchmarks can be reference points for measuring and rewarding good performance. In the context of this report, building owners can use benchmarking and baselining to measure the effect on overall building performance of new processes, practices, and behaviors driven by submetering data.

Because buildings are of many varied types and uses, it is essential that any benchmarking activities focused on identifying best-in-class submetering performance explicitly recognize and address these differences. More comprehensive performance metrics such as ENERGY STAR® Performance Ratings[37] will differ across building types, and include climate zone and building characteristics such as operating hours, building size, occupancy, and number of computers.

Baseline values, or average levels of performance, are also a valuable reference point, as they provide a measure of the improvement opportunity achievable through benchmarking. A key source of baseline values is the Commercial Buildings Energy Consumption Survey (CBECS).[38] CBECS tracks energy consumption and expenditures for 12 building types and five climatic zones, and employs a weighted sample of buildings that collectively represents the entire stock of commercial buildings in the United States. Public use microdata files are also available via the CBECS Web site; these files allow more detailed cross tabulations than are provided in the various CBECS reports.

For energy use, a commonly used metric is *energy use intensity* (EUI), typically defined as the overall energy consumption of a building divided by the total floor area of the building. It is important to understand that judging a building's relative performance as indicated by the EUI depends significantly on the location and type of building. A recent study on building energy use intensities found that metrics could vary based on the performance criteria used and highlights the importance of choosing the right performance metric[39] for the application. Oak Ridge National Laboratory has compiled data on 24 building types at various locations throughout the United States to which a building can be compared,[40] while EPA has created a list of 26 national average EUIs for buildings, which are currently not eligible to receive an ENERGY STAR energy performance score, to use as reference energy performance targets. Through Portfolio Manager, EPA also provides energy performance scores that account for the impact of weather variations as well as changes in key physical and operating characteristics for 14 building types.[41]

For water benchmarking, the term water use refers to water used for any practical purpose, including consumption. Water use benchmarks are either absolute or relative: absolute benchmarks define a quantity of water used; relative benchmarks represent a unit-less ratio of water used divided by a target value quantity. The most common bulk metric for water use is "per capita use" measured in gallons per person per day. The common water metric specific to water devices is gallons per hour (gph), and for reporting Federal water use mandated by EO 13514, gallons per square foot is used.

Challenges exist when using metering data to measure conservation impacts due to the aggregated use of water in the building that captures "other than efficiency" effects. And, when evaluating the water efficiency of a device relative to an absolute water use benchmark, the measurement must not only include the flow rate (gph) of the device but also the effectiveness of the device to achieve the desired results—a showerhead with a 50 percent lower flow rate that requires twice as long to rinse is no more efficient that the full flow-rate device.

From the ENERGY STAR Guidelines for Energy Management[42], the process for performance assessment is outlined below:

Baselining and Benchmarking	
Gather and track data	Collect resource use information and document data over time.
Baselining and Benchmarking	
Establish baselines	Determine the starting point from which to measure progress.
Benchmark	Compare the performance of your facilities to each other, to peers and competitors, and over time to prioritize which facilities to focus on for improvements.
Analysis and Evaluation	
Analyze	Understand your resource use patterns and trends.
Technical assessments and audits	Evaluate the operating performance of facility systems and equipment to determine improvement potential.

4.3.2. Communication Networks and Data Storage Requirements

Regardless of the meter type, once submetering data is collected it must be transmitted via a communication network to be processed, stored, and used. The type of submetering plan with its requisite data requirements will define the communication network appropriate for the application. Current options for automated metering communications include phone modem, local area network (LAN), building automation system, radio frequency (RF), and wireless network. Submetering devices connected to existing building automation system networks provide an easy and cost-effective communication alternative by taking advantage of a site's previous investment in network infrastructure.

As mentioned previously, when data communication networks are connected to existing, secured data networks, the communication protocols should include appropriate levels of information assurance and security[43]. To achieve these network security requirements, the control protocols have been extended with a set of network layer security messages that provide data confidentiality and integrity, device authentication, data hiding, and user authentication. For wireless communication systems, see also standards IEEE 802.11 (WiFi) and IEEE 802.15.4 (appropriate for low data rate applications) for more information.

For continuous-performance monitoring systems, data storage and database systems should also meet common security and integrity requirements. The California Energy Commission's Public Interest Energy Research Program (PIER) and the Building Technologies Program of the U.S. Department of Energy have developed data specifications appropriate for energy management systems.[44]

Data Storage Specification Considerations

- Data will be stored in a structured query language (SQL)-compliant database format or time-series format. The minimum requirement is an SQL server or equivalent.
- The database will allow other application programs to read and access the data with appropriate password protection while the database is running. The database will not need to be shut down in order to access or add data.
- Trend data will be archived in a database from field equipment in time intervals of no less than once per day.

- Storage on the field equipment will be reset once data are exported to allow for trending if communication is disrupted. Data will be uploaded once communication is reestablished.
- Blank or null values in the database will be replaced with actual data.
- Calculations and other metrics will be updated once controller data are uploaded. This overall system update to check for new data should be automated to run once a day.
- All data will be stored in a database file format for direct use by third-party application programs (e.g., Energy Information System [EIS]).
- Sufficient data storage capacity will comprise at least two years of data for all data points. In addition, storage capacity will allow for compression of one year of data for historic trends and archiving.
- Time stamps will be collected on all data. The time stamp, depending on system architecture, will be captured by the field controller or system controller and directed to the database archive.
- Exported data will contain no duplicate records or duplicate time stamps in output files. Each date/ time stamp for a specific point will be unique. The export query will be for a specific point or multiple points in a defined group.
- Date/time fields will be in a single column in a format automatically recognized by common spreadsheet, database software tools, or EIS.
- The data will be fully contained in a single file or table for each point. Data will not span multiple files or database tables. Users can have the option to modify the start and end dates for exporting files depending on third-party program requirements to evaluate the data.

4.3.3. Submetering Information Systems

Once installed, meters, submeters, and automated building control systems can generate large volumes of data that must be stored, analyzed, and archived. Metered energy data is analyzed at regular intervals (e.g., every 15 minutes is standard in the U.S.) and may incorporate other relevant information:[45]

- Account or location
- Usage points (e.g., lighting, HVAC)

- Kilowatt-hours of consumption
- Highest kilowatt demand
- Voltage
- Real and reactive power
- Hourly weather information (temperature, humidity)
- Dollar values of consumption, peak charge, access charge, taxes, and delivery charges
- Calculated values for load factor, per square foot or per square meter use
- Heating or cooling degree days

For water distribution systems, metered data may include:

- Water flow rate
- Peak flow rate
- Pressure
- Temperature
- Hourly weather information (temperature, humidity)
- Location

This information alone provides no significant insights into building performance unless it is integrated with software analysis capabilities. For energy systems, an Energy Information System (EIS) refers to the broad class of analysis tools ranging from spreadsheets on local computers to integrated systems with analysis capabilities, trend identification, report generation, and graphical displays. A building's advanced meter, submeters, and building automation systems can all provide data to the EIS.

Formally, EIS refers to software, data acquisition hardware, and communication systems administered by a company, partnership, or collective to provide energy information to commercial building energy managers, facility managers, financial managers, and electric utilities.[46] Appendix D contains a table of current EIS products. These systems provide a range of data acquisition, analysis, diagnostic and targeting, and reporting capabilities. The complexity of each system depends on the application and the desired level of monitoring. The text box below highlights many of the features to be considered when selecting an EIS.

Moreover, installing, managing, and communicating results from EIS systems requires technical staff and equipment, which can consume additional staff resources. A number of companies offer EIS services for a fee. These services allow the building operator to select the level of analysis granularity in reporting, and they provide expertise in diagnosing problematic, failing, or inefficient building systems.

Scoping EIS Capabilities

- Definition of facility objectives—How will the EIS will be used? How will it improve resource efficiency and its contribution to the financial payback of the overall metering system?
- EIS integration with other IT systems—What existing infrastructure can be used? A survey of existing facility software, hardware, communications, meters, and capabilities is recommended before the EIS decision is made.
- Key reporting outputs—How will the outputs of this system be used to fulfill the objectives of the facility and its management? How will the outputs be made available (paper copy, email reporting, Web-based reporting), and what summary options are available?
- Data-collection needs—What are the data-collection requirements for achieving the key reporting outputs? Are they available with existing metering elements?
- Data analytics—What analysis, statistical, and/or regression routines are needed to transform the data into the information desired? Are the chosen analytics capable of handling the size, frequency, and complexity of the anticipated data?
- Management support—Has a budget been developed to address expenses, such as ongoing training, periodic testing, technical assistance, and troubleshooting?

Taken in part from: Hooke, J.H., B.J. Landry, and D. Hart. *Energy Management Information Systems: Achieving Improved Energy Efficiency*. Office of Energy Efficiency, Natural Resources Canada.

4.3.4. Submetering Analysis and Graphical Interfaces

Analysis and graphical display software provide a powerful and convenient means for building owners and operators to assimilate building performance information and to identify opportunities for improving resource use. Dashboards with features and interfaces that are tailored to the needs of building owners, operators, and occupants are currently under development. Dashboards exist for both energy and water use applications.

Distinct from conventional Energy Management and Control Systems (EMCS), these dashboards typically do not provide detailed system operation. Instead, they are designed to display trends and anomalies, and to educate a broad range of building stakeholders about the ecological implications of building performance and user behavior. Many of these products include real-time information displays and allow users to view data using a number of different metrics, such as energy units, utility costs, or carbon emission equivalents.[47]

EIS Graphical Outputs

- Daily profile—Displays time-series daily load profiles with time, in intervals of an hour or less, along the horizontal axis and load along the vertical axis.
- Day overlay—Displays multiple daily profiles on a single 24-hour time-series graph.
- Multi-point overlay—Allows viewing of multiple time-series data points on the same graph.
- 3-D surface chart—Displays the time of day, date, and variable for study.
- Calendar profile—Presents up to an entire month of consumption profiles on a single screen as one long-time series.
- X-Y scatter plots—Allow visualizing of correlations between two variables.

Taken in part from: Federal Energy Management Program, "Metering Best Practices: A Guide to Achieving Utility Resource Efficiency," October 2007.

Visualization products by a number of companies provide bundled suites of analysis software and visualization tools. These applications include energy dashboards for both the end user and the facility manager and provide detailed usage metrics, trends, and outputs for management and carbon reporting requirements. They also include real-time load profiles and baseline comparison profiles that correct for weather and other external conditions, so that occupants and building managers can accurately determine the impact of energy conservation initiatives.[48] Recent software analysis offerings can be found in Appendix D.

A report from Lawrence Berkeley National Laboratory, entitled "Building Energy Information Systems: State of the Technology and User Case Studies," presents the following summary of the state of EIS technology:[49]

Business models (General)

- EIS are most commonly offered through an Application Service Provider (ASP) with no hardware or optional hardware based on client needs.
- Optional or bundled services are nearly universally offered.

Display and visualization

- Features have converged to a near common set. Data can be viewed over user-defined intervals of time, trended variables can be aggregated into totals, and the user can overlay multiple datasets on a single plot.
- X-Y scatter plotting is offered in only half of today's EIS solutions.

Energy analysis

- Two-thirds of the EIS feature greenhouse gas analysis or provide custom or configurable options to do so. Most apply a simple energy/carbon dioxide (CO_2) relationship, but almost half account for regional differences in generation or other standards.
- Nearly every EIS permits the user to quantify an energy consumption baseline; however, weather normalization is rare.
- Every tool that was evaluated supports (or will soon support) multi-site benchmarking. Distinguishing aspects include:
- Composition of the comparative cohort: buildings within the user's enterprise; comparison to buildings from the vendor's database; or less commonly, national data sets.
- Display of results: static reports versus dynamically accessible functions; results depicted in tables, plots, or charts.

Advanced analysis

- About three-quarters of the EIS address data quality, and they do so via three principal means: flagging or summative reporting, cleansing and/or correction, and linking to external or third-party software packages.

- Anomaly detection is typically trend-based and accomplished by identifying departures from normal energy consumption patterns.
- More than half of the EIS forecast near-future loads, usually by coupling historic trends and weather data; very few provide model-based capabilities.
- The large majority of the EIS accommodate some form of measurement and verification (M&V) or the ability to track the impact of operational changes.

Financial analysis

- Energy costing is supported in nearly all of the EIS, and more than half have implemented model-or tariff-based calculations.

Demand response

- Demand response (DR) capabilities have advanced since early 2000 and have converged to a common set of features.
- Automated response to DR signals is supported in all but three of the DR systems that were characterized.
- Remote control and management
- Just over half of the EIS surveyed report the ability to control according to a program, and just under half report internet-capable direct remote control.

Dozens of EIS are commercially available, yet public domain information is often vague, and demonstration software may not be available. In addition, a lack of common terminology across vendors makes it difficult to discern tool functionality or hardware requirements, or to compare the effectiveness of products.[50] Between 60 and 80 percent of the energy used in office buildings is consumed by tenants within their spaces, yet no widely available tools currently exist to help tenants understand their energy consumption or compare it against their peer group.[51] EPA's Portfolio Manager tool, for instance, is widely used for establishing building-level energy performance and ranking, but there is no comparable tool that allows tenants to do the same. Along with providing tenants their own consumption profiles and data, the EIS should offer an energy utilization index—a comparative ranking of tenants against their peer group based on their energy intensity.

5. Summary and Recommendations

Submetering energy and water usage in buildings represents an important first step toward quantifying over time the point sources that can drive resource efficiency and conservation activities. Submetering is an enabling technology with significant potential value in new building design or the retrofitting of existing buildings. Submetering alone has no direct impact on resource use, but it provides building owners with real-time system energy and water performance data, highlights performance variations over time or in comparison to other buildings, can feed building automation systems, and establishes a basis for behavioral and operational changes by building operators and occupants.

While the cost effectiveness and performance of submetering will vary by building type, building use, climatic considerations, and other factors, numerous case studies provide evidence that the return on investment (ROI) can be significant.

Further, submetering provides the necessary infrastructure for more advanced conservation and efficiency techniques such as monitor-based commissioning and autonomic building operation.

This report provides the building community with an understanding of the full complexity and potential benefits of using submeters to improve building performance. The role of submetering is expected to increase over time as building performance requirements become increasingly stringent. To meet these future requirements, the NSTC Subcommittee on Buildings Research and Development makes the following recommendations (Table 2).

Table 2. BTRD Recommendations for Submetering

Recommendation 1: As part of overall building design and retrofit projects, building owners should evaluate the economic and technical feasibility of submetering using life cycle cost analysis for the economic justification and business-case development.

Recommendation 2: When evaluating submetering opportunities, design teams should collaborate with building owners, operators, and occupants, and design, engineering, and information technology professionals to identify data needs and appropriate submetering systems and configurations.

Recommendation 3: Submetering systems should, when possible, be integrated into building automation systems and used to identify building system impairments or dysfunction, to improve operational procedures, drive behavioral changes that improve building energy efficiency and conservation, and augment building-side automation systems that interface with the Smart Grid.

Recommendation 4: Submetering information and feedback mechanisms should be tailored to intended users to effect operational and behavioral change in building operators and occupants. Information and feedback should be of an appropriate form and frequency and should focus on useful breakdowns of resource use; include historical data when appropriate to facilitate comparisons with past use; and, when possible, suggest specific actions that can be taken to reduce consumption.

Recommendation 5: Submetering systems should be operated as production-quality building management systems with high levels of reliability, information assurance and security, established data acquisition/ archival procedures, and periodic system maintenance requirements.

Next Steps: The Subcommittee for Buildings Technology Research and Development is considering two possible activities to follow the publication of this report. These activities will augment the information provided in this report:

a) Development of a checklist or decision flow chart for building owners/operators/managers that guides them through decisions as to when submetering is appropriate and supporting details such as number of submeters, types of meters, types of data collection systems, data analysis tools, etc. This supplemental information will allow for a more rigorous cost-benefit analysis than the case studies provided.
b) Establishment of a stakeholders process to build an open-source platform for voluntary collecting of energy consumption data and metadata associated with Federal and privately owned commercial buildings, and for enabling open access to aggregated data for building performance benchmarking and other applications.

Appendix A: Summary of Regulations for Benchmarking and Metering

While there are relatively few legal requirements or guidance specifically dealing with submetering building resource use, there are many related efforts underway at Federal, state, and local levels in both the public and private sectors. These efforts can be classified into three main types: Federal benchmarking and metering activities, state and local benchmarking and metering activities, and private sector activities, including code development. This appendix provides an overview of some of these activities relevant to building resource metering and submetering.

A.1. Federal Benchmarking and Metering Activities

Statutory Requirements

The Federal government has metering requirements for Federal-owned buildings. Section 103 of the Energy Policy Act of 2005 (EPAct 2005) requires all Federal agencies to install metering and advanced metering *to the maximum extent practicable* by October 1, 2012, to ensure efficient energy use and reduce the cost of electricity used in Federal facilities. The Energy Independence and Security Act of 2007 (EISA 2007) adds a number of additional requirements relevant to submetering. Section 432 of EISA 2007 establishes a framework for facility project management and benchmarking. Agencies must identify all "covered facilities" that constitute at least 75 percent of the agency's facility energy use. Each covered facility must designate an energy manager who will be responsible for:

- Completing comprehensive energy and water evaluations of 25 percent of covered facilities each year, so each facility is evaluated at least once every four years.
- Following up on implemented measures, including fully commissioning equipment, implementing O&M plans, and measuring and verifying energy and water savings.
- Using a DOE Web application to certify and track compliance for energy and water evaluations, project implementation and follow-up measures, and estimated cost and savings. The results will be available to Congress, other Federal agencies, and the public, with

some specific data exempted from disclosure for national security purposes.

- Entering energy use data for each metered building into a benchmarking system, such as ENERGY STAR Portfolio Manager.

EISA sections 323 and 435 add energy performance requirements for new construction and leased spaces.

Advanced meters (including submetering systems) are critical catalysts for compliance with these requirements because they enable identification of life cycle cost-effective measures, provide benchmark data, and support monitoring and verification of energy performance. Advanced meters or metering devices must provide data at least daily and measure the electricity consumption at least hourly. Each Federal agency is required to submit to DOE an implementation plan identifying the personnel responsible for achieving metering requirements as well as any determination that advanced meters or metering systems are not practicable in the agency's specific situation. FEMP has issued guidance for this requirement in its *Guidance for Electric Metering in Federal Buildings,*[52] which includes an overview of electric metering technologies and approaches, information about metering cost-effectiveness, and a section on financing metering projects. FEMP has also developed a best practices guide entitled *Metering Best Practices: A Guide to Achieving Utility Resource Efficiency.*[53]

Executive Order 13514

Presidential Executive Order 13514,[54] signed by President Obama in October 2009, establishes 23 separate goals for Federal facilities to reduce greenhouse gases, increase the use of renewable resources, and pursue cost-effective, innovative strategies to minimize consumption of energy, water, and materials. This directive requires Federal agencies to accurately and consistently quantify and account for greenhouse gas emissions using accepted greenhouse gas accounting and reporting principles. Agency efforts and outcomes in implementing EO 13514 must be transparent and disclosed on publicly available Federal Web sites. The installation and use of meters and submeters will be an essential component for Federal agencies to meet the requirements of this order and the requirements for EPAct 2005 and EISA 2007. This directive also requires that Federal facilities improve water use efficiency and management by reducing potable water consumption intensity by implementing water management strategies including water-efficient and low-flow fixtures and efficient cooling towers.

U.S. Department of Defense Instruction 4170.11

The U.S. Department of Defense (DoD) accounted for approximately two-thirds of the total energy used by Federal buildings in 2005. DoD Instruction 4170.11, Installation Energy Management (November 22, 2005 and revised on December 11, 2009), requires DoD to meter electricity, natural gas, and water at appropriate facilities (those where "metering would be cost-effective and practical as a management enhancement tool") and to meter steam at steam plants.

A.2. State and Local Benchmarking and Metering Activities

A number of U.S. states and cities are also developing building-scale data collection programs in support of sustainability initiatives. California, Washington,[55] and four major metropolitan areas (New York City,[56] Austin,[57] Seattle,[58] and Washington, DC[59]) have mandated benchmarking of energy use (and in one case, water use) for large classes of buildings. In addition, major cities such as Atlanta, Boston, Baltimore, Miami, Portland, and San Francisco are considering proposals to mandate benchmarking for commercial buildings. Table A1 below provides an overview of these activities, including which types of buildings are benchmarked, whether utility data is being used, whether benchmarks are disclosed and how, and the resources being measured.

These programs will generate large amounts of data. For example, New York City's mandate calls for annual aggregate energy and water use data to be benchmarked for approximately 20,000 buildings and eventually published as part of the city's database of property assessments.

Table A1. Summary of State and Local Benchmarking Requirements[60]

Jurisdiction	Public Buildings	Private Buildings	Disclosure	Utility Data	Building Requirement	Resource Measured
California	√	√	Transactional	√		Energy
Washington	√	√	Transactional	√		Energy
Michigan	√					Energy
Ohio	√					Energy
Hawaii	√				Public > 5,000 ft^2 or > 8,000 kWh	Energy
Washington, DC	√	√	Annual		Private > 50,000 ft^2 Public > 10,000 ft^2	Energy

Table A1. (Continued)

Jurisdiction	Public Buildings	Private Buildings	Disclosure	Utility Data	Building Requirement	Resource Measured
Austin, TX	√	√	Transactional		Commercial > 10 yr old	Energy
Denver, CO	√					Energy
West Chester, PA	√	√				Energy
Seattle, WA	√	√	Transactional	√	Commercial > 50,000 ft^2 Multi-unit res > 5 units	Energy
New York, NY	√	√	Annual		Commercial > 50,000 ft^2 Multi-unit residential Municipal > 10,000 ft^2	Energy, Water

The primary objective of building submetering is to identify systems or areas of high resource use and make changes to those systems to conserve resources. One strategy used by cities leverages this objective by requiring public disclosure of billing data, with the idea that disclosure of building consumption is useful for performance verification by potential buyers and renters, adding transparency to real estate and rental property evaluations. For potential owners or occupants, knowing a building's energy and water usage can be useful information when comparing similar buildings. If all else is equal, why not choose the more energy- or water-efficient building?

There are a number of examples of established public disclosure programs, and probably the oldest and most well-known is ENERGY STAR.[61] The ENERGY STAR label or rating is conferred to buildings in the lowest 25th percentile of energy consumption for each building type, as reported in the most recent Commercial Buildings Energy Consumption Survey (CBECS).[62] Currently in pilot phase, the Building Energy Quotient program[63] from ASHRAE (American Society of Heating, Refrigerating, and Air-Conditioning Engineers) is another example of a labeling program that may be used to satisfy mandatory public disclosure requirements. All of these disclosure requirements and labeling programs harness the power of the marketplace and the desire of potential tenants and buyers of energy-efficient buildings.

ENERGY STAR Program

ENERGY STAR, EPA's voluntary partnership program, offers a proven energy management strategy that helps in measuring current energy performance, setting goals, tracking savings, and rewarding improvements. Building owners and operators can benchmark any commercial facility in EPA's Portfolio Manager tool to track energy, water, environmental, and financial performance over time. EPA's 1-100 energy performance score is also available for 14 building types, including offices, schools, hotels, retail stores, and many more. For most of these building types, statistically representative models are used within Portfolio Manager to compare a building against similar buildings from a national survey conducted by the Department of Energy's Energy Information Administration. This national survey, known as the Commercial Building Energy Consumption Survey (CBECS) is conducted every four years, and gathers data on building characteristics and energy use from thousands of buildings across the United States. The models for hospitals, data centers, and wastewater treatment plants are based on robust industry surveys that gather statistically representative data on specific operating parameters and characteristics. The score is based on source energy and accounts for the impact of weather variations as well as changes in key physical and operating characteristics of each building. Buildings rating 75 or greater may qualify for the ENERGY STAR label.

For more information, see http://www.energystar.gov/index.cfm?c=evaluate_performance.bus_portfoliomanager#rate

A.3. Model Code Requirements

In addition to government-led initiatives, some building energy codes have mandated metering and submetering. Submetering of individual tenants in multiple-tenant buildings was once a requirement in the national model energy codes. ANSI/ASHRAE/IESNA Standard 90.1-1989 Section 5.4.1.2 stated, "In multiple tenant buildings, each tenant shall be individually metered. Provisions to permit check metering of the tenant load shall be provided for those tenants having a connected load of 100 kVA or more. HVAC or solar water heating (SWH) systems shared by tenants in common need not meet this tenant check metering requirement but shall be separately metered as required." The term check metering was defined as "measurement instrumentation for the supplementary monitoring of energy consumption (electric, gas, oil, etc.) to isolate various categories of energy use to permit conservation and control, in addition to the revenue metering furnished by the utility."

As a method of providing information to tenants so they could make informed decisions on energy conservation, this requirement was appropriate and ahead of its time. Unfortunately, it was deleted from later versions of Standard 90.1 and was never found in the model energy codes or the International Energy Conservation Code. Some state and local codes, however, picked up this requirement or a variant of it (e.g., the Chicago Energy Code).[64]

There has been renewed interest in metering and submetering in the energy code development communities. There were proposals to reintroduce metering requirements in Standard 90.1, but the proposals did not pass in time to be included in the 2010 version. ASHRAE/USGBC/IES Standard 189.1 2009 [65] contains sections that require metering of energy and water usage (see Section 6.3.3 for water consumption management and Section 7.3.3 for energy consumption management). The draft International Green Construction Code (IGCC)[66] contains similar provisions (see Section 604 for energy metering, monitoring, and reporting; Section 604.5 for energy load-type submetering; Section 611.5 for renewable energy system performance monitoring and metering; and Section 705.2 for water metering).

APPENDIX B: SUMMARY OF BEHAVIORAL RESPONSES TO ENERGY FEEDBACK

Residential sector research of multi-tenant or multi-space buildings shows that increasing consumers' awareness of their resource consumption patterns can lead to conservation. The installation of smart meters or submeters, if coupled with appropriate data, networks, analysis tools, and strategies, can provide stronger feedback mechanisms to help end users monitor, understand, and reduce their resource use.

Residential sector research has also focused on consumers' behavioral responses to feedback on energy consumption. There are a number of proven methods to inform building occupants and operators about their energy use—some methods are specific to end users, while others are best suited to building operators. None requires the use of a smart meter or submeter, although these devices can be effectively used to provide ongoing month-to-month performance data.

Feedback Mechanisms

Direct feedback is delivered immediately from a meter or an associated display monitor. This type of feedback ties an action to its energy cost by showing the cost immediately after the action. However, this raw usage data may also lead to information overload and overwhelm a building occupant or operator.

Indirect feedback has been processed in some way before reaching the consumer, as in a monthly utility bill. Improved forms of indirect feedback, including more frequent mailings or more salient billing (e.g., illustrative graphs) have been effective in significantly reducing energy use. Indirect feedback may also incorporate historical analysis or other creative metrics; however, the chief disadvantage is the time lag between action and feedback.

Disaggregated feedback uses the unique signatures of smart appliances to identify costs by end use. (The algorithmic approach is sometimes referred to as non-intrusive appliance load monitoring.) This can help consumers identify the most significant sources of resource use. On the other hand, the technology to perform disaggregation can be expensive and may not always be cost-effective.

Historical feedback compares present usage to usage in a prior time period (e.g., the same month in the previous year). Historical feedback allows consumers to benchmark their progress, though due to climate variability, it may not always provide a reliable comparison.

Comparative feedback presents a user's consumption with that of a comparable set of users (e.g., grouped by building, organization, or zip code) in an attempt to produce conservation through social or operational norms. One criticism of comparative feedback is the possibility of triggering a boomerang effect in which lower-than-average consumers increase resource use to match the average usage rate of the peer group; another is the possibility of antagonizing users who resent comparison.

Digital feedback uses computer and electronic displays to present multiple or interactive viewing options for feedback presented to the consumer.

Behavioral approaches also can help end users understand and reduce their energy and water use. These approaches include tailored information, goal setting, rewards, and time-of-use pricing mechanisms to reduce peak load. All have been shown to be effective in some circumstances and ineffective in others; they are often used in tandem with feedback in the objective of producing synergistic effects. Literature reviews from the last 10 years[67] show that feedback mechanisms result in energy reductions ranging from 0 to 20 percent. However, the quality of the studies is inconsistent. Many studies are too short in duration, too small in sample size, or too uncontrolled to allow for meta-analysis or extrapolation to predict long-term effectiveness. Due to this variability, more examination is required to establish which types of feedback mechanisms, at which time intervals and through which media, will most substantially contribute to energy conservation in specific settings. Nevertheless, some consensus has emerged regarding the kinds of feedback interventions that produce the most energy conservation:

Direct feedback—The reviews generally concurred that direct feedback is more effective than indirect feedback, although indirect feedback leads to substantial savings in heating loads (which tend to manifest over longer periods).

Frequency—More frequent feedback tends to be more effective. In particular, feedback on at least a daily basis improves conservation. More frequent feedback may also lead to more persistent savings, even after the feedback has been discontinued, though this result is less well established.

Disaggregation—Detailed appliance-specific breakdowns of energy use tend to improve conservation. However, it is not as clear whether disaggregation is more effective if the data are presented via the appliance itself or remotely (e.g., from a computer).

Digitization—Computerized feedback programs, especially those with multiple and/or interactive viewing options, tend to be the most effective. Systems that allow the consumer to pick the most salient view of energy use appear to increase engagement.

Historical and comparative feedback—Studies, which included interviews, found that households tend to like detailed feedback that includes historical and comparative feedback.[68] However, they also found no significant difference between the effectiveness of historical and comparative feedback used independently, and the overall effectiveness of both methods when taken together.

Advice—Energy feedback containing personalized advice about specific actions to reduce consumption tends to enhance energy conservation. Yet advice alone is rarely effective—it tends to increase knowledge of energy-efficient behaviors but does not typically affect actions.[69]

Commercial buildings—In residential settings, the desire to save money on utility bills motivates individuals to conserve. In office settings, a similar motivation may not exist. Relatively little is understood on how to encourage conservation in this context.

Feedback design—Optimal location, content, units, and timescale of feedback require additional understanding. Only a few studies have examined the difference between, for example, giving feedback in dollars versus kilowatt-hours. In addition, very little is known about the kinds of electronic media that are most effective at engaging consumers. This could be an important factor in office settings inundated with electronic devices and displays.

Granularity—Research has typically shown that more granular feedback is more effective. However, there have been few studies comparing different timescales of feedback, and their wide variability in quality makes inter-study comparison difficult.

Source of data—In the residential sector, the source of information—e.g., a utility versus a third-party group—can affect its reception.[70] More work is needed in both residential and commercial settings to better understand what sources of feedback are likely to be heeded most often.

Combination with other techniques—Many studies and operational feedback-based conservation campaigns combine feedback with non-feedback elements, such as goal setting. However, there is not yet a systematic understanding of the synergistic or antagonistic interactions between behavioral interventions.

Long-term effects—Most studies last on average a few months, although some have been as short as two weeks or as long as two years. As a result, relatively little is known about the long-term effects of feedback, or about the lasting effects of feedback once it is removed.

APPENDIX C: SUBMETERING EQUIPMENT, CONFIGURATIONS, AND COSTS

The images and tables provided courtesy of Square D, a brand of North American Operating Division of Schneider Electric, Inc., for use in this report. The data represents a fictitious sample implementation and associated costs.

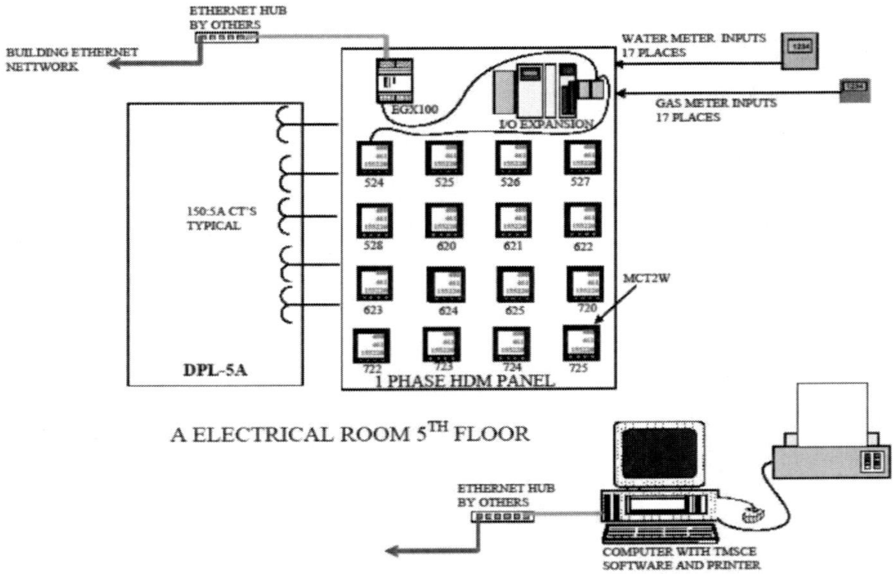

Qty	*Description*	*List Ea.*	*Total List*
	Gas & Water Metering Kits - Pulse Inputs (Field Installed in HDM Box)		
4	HDM Ethernet and I/O Expansion Kit: 1-EGX100, PS & Twido I/O 120/240V - 24VDC P/S (44 Digital Inputs)	$ 2,280	$ 9,120
	NEMA 1, Single Phase		
3	HDM, 120/208/240V, 1Ph, 1 Meter Box, 1 meter installed - PM210	$ 400.00	$ 1,200
	PM210 Tenant Submetering (240 Volt Max, 1 Phase)		
2	HDM, 240V Max, 1Ph, 16 Meter Box, 14 meter installed - PM210	$ 9,271	$ 18,542
2	HDM, 240V Max, 1Ph, 16 Meter Box, 16 meter installed - PM210	$ 10,342	$ 20,684
	Accessories		
4	2-Wire Modbus Terminator	$ 84	$ 336
	Tenant Metering Software		
1	Tenant Metering Software Commercial Edition	$ 5,880	$ 5,880
	PMO-ES Standardized PC Hardware		
1	OptiPlex 755 Minitower, Core 2 Duo E6550/2.33GHz 4M VT 1333FSB, 2GB DDR2 Non-ECC SDRAM, 800MHz (2DIMM), 160GB High Reliability SATA 3.0Gb/s and 8MB Data Burst Cache, Integrated Video (GMA3100), ONBOARD NIC, Windows XP SP2 Pro, 48X32 CDRW/DVD Combo, PS2 Serial Port Adapter, USB KB / Mouse, Energy Smart/Energy Star, vPro Secure Advanced Hardware Enabled Systems Management	$ 1,606	$ 1,606
1	Dell Ultra Sharp, Flat Panel Monitor, 19" LCD, 1280x1024 Max Pixel Resolution, Digital DVI-D and Analog Inputs, Adjustable Height Stand w/Swivel & Tilt, 4 USB 2.0 Ports	$ 662	$ 662
1	Color Laser Printer, Letter and Legal Size Paper, 600x600 DPI, USB 2.0	$ 1,531	$ 1,531
	HDM CT's		
120	150 Amp HDM Solid Core Current Transformer 1.13" Window Size	$ 35	$ 4,200
	Cold Water Meters		
65	3/4" X 7.5" LowLead C700 with Glass Digital Register	$ 213	$ 13,872
	Gas Meters		
65	G4 Gas Meter with 1 pulse per cubic foot, Sprague #1 connections, non-compensated	$ 398	$ 25,870
	Startup & Commissioning Services: PowerLogic Tier A Project		
1	Onsite Startup/Commissioning work per PMO Engineering Services Standard Scope of Work.	$ 1,778	$ 10,668
1	Documentation package that includes preliminary/as-built system drawing, as required, and project files.	$ 833	$ 833
1	Onsite Customer System Training for ES Projects Provided solutions - brief overview of products and system should be included in Startup & Commissioning quote.	$ 1,778	$ 1,778
			$ 116,782

76	**Maintenance and Labor Costs Per Year**											
77												
78	**Item**	**A.Price /Item**		**B.Labor**		**C.Admin. /Item**		**D.No. Items**	**E.Total Equip.**	**F.Total Labor**	**G.Total Admin.**	**H.Total**
79				**Field Engineer I**		**Project CoPI**						
80		%	Price	Hrs	$/Hr	Hrs	$/Hr		(=A X D)	(=B X D)	(=C X D)	(=E+F+G)
81	Logger C-180	10%	$279.50	2		0.2		1	$280	$116	$23	$418
82	Logger C-140	10%	$184.50	2		0.2		0	$0	$0	$0	$0
83	Logger C-120	10%	$124.50	2		0.2		0	$0	$0	$0	$0
84	Logger B-80/Datamate	10%	$74.50	2		0.2		0	$0	$0	$0	$0
85	Portable Logger A	10%	$0.00	2		0.2		0	$0	$0	$0	$0
86	Portable Logger B	10%	$0.00	2		0.2		0	$0	$0	$0	$0
87	Portable Logger C	10%	$0.00	2		0.2		0	$0	$0	$0	$0
88	DK Btu Meter	10%	$50.00	2		0.2		0	$0	$0	$0	$0
89	Temperature Sensors	30%	$21.00	2		0.2		4	$84	$462	$92	$638
90	Flow Meter (DI)	30%	$180.00	4		0.2		2	$360	$462	$46	$868
91	Flow Meter (FR)	30%	$150.00	4		0.2		0	$0	$0	$0	$0
92	Flow Meter (Onicon)	30%	$255.00	4		0.2		0	$0	$0	$0	$0
93	Flow Meter Calibration	0%	$0.00	2	$58	0.2	$116	1	$0	$116	$23	$139
94	Condensate Meter	30%	$540.00	4		0.2		0	$0	$0	$0	$0
95	CTs	10%	$15.00	1		0.1		6	$90	$347	$69	$506
96	PTs	10%	$17.50	1		0.1		1	$18	$58	$12	$87
97	Watt Hour Transducer	10%	$39.40	1		0.1		0	$0	$0	$0	$0
98	Phone	10%	$0.70	0.5		0.1		1	$1	$29	$12	$41
99	Conduit	0%	$0.00	0.1		0.1		1	$0	$6	$12	$17
100	Handheld Instrument A	20%	$0.00	0.2		0.1		0	$0	$0	$0	$0
101	Handheld Instrument B	20%	$0.00	0.2		0.1		0	$0	$0	$0	$0
102	Handheld Instrument C	20%	$0.00	0.2		0.1		0	$0	$0	$0	$0
103	other	0%	$0.00	0		0		0	$0	$0	$0	$0
104	other	0%	$0.00	0		0		0	$0	$0	$0	$0
105												
106	**Total Costs**								$832	$1,594	$289	$2,714

The image below was developed by the Energy Systems Laboratory U.S. Army Construction Engineering Research Laboratory as part of *U.S. Army Measurement and Verification (M&V) Costing Toolkit,* published in December 2003 for the U.S. Army Construct on Engineering Research Laboratory.

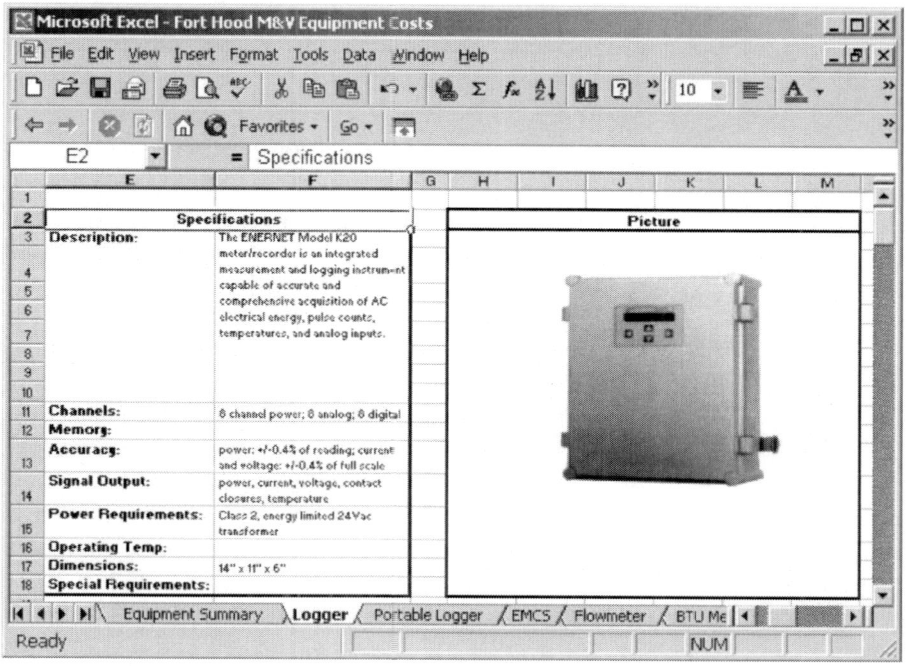

Cost estimate for a electricity submeter installation on a college campus based on six submeters in three separate locations (data provided courtesy of Kapadia Energy Services.)

Item Description	Quantity	Unit Material Cost	Unit Labor Cost	Total Cost
Electricity submeter with demand display and pulse device	6	$1,200	$450	$9,900
Current transducers	18	$300	$320	$11,160
16 point PLC reading board	3	$1,400	$600	$6,000
Terminal Interrogation Module w/modem	3	$2,400	$500	$8,700
Windows-based meter reading software	1	$3,500	$200	$3,700
Supervision of installation and setup	1		$1,500	$1,500
Total Cost:				$40,960

APPENDIX D: RECENT EIS SOFTWARE PRODUCTS

Vendor	Product Name	URL
Abraxas Energy Consulting	EnergyCAP	https://www.abraxasenergy.com/energycap/
Abraxas Energy Consulting	Metrix4 Utility Accounting System	http://www.abraxasenergy.com/metrix4/
Activelogix	Periscope	http://www.activelogix.com/about_periscope.asp
Advantage IQ	Energy Services	http://www.advantageiq.com/WhatWeDo/EnergyServices/tabid/174/Default.aspx
Agilewaves	Building Optimization System	http://www.agilewaves.com/resource-monitor-commercial/
BITHENERGY	Utility Data Analysis Software	http://www.bithenergy.com/Utility-Data-Analysis.html
BizEE	Energy Lens	http://www.energylens.com/
Burton Energy Group	Management Reports	http://www.burtonenergygroup.com/utility-information-management.html

Vendor	Product Name	URL
E Source	E Source Facility Metrics	http://www.energy-
EFT Energy	Energy Manager V5.2	http://www.eft-energy.
Energard Technologies	Envision	http://www.energard.com/page3.htm
enerGXpert	enerGXpert	http://www.energxpert.com/
Energy Essentials	Stark RT & Stark Essentials	http://www.starkna.net/about.htm
ENERGY STAR	Portfolio Manager	http://www.energystar.gov/index.cfm?c=evaluate_performance. bus_portfoliomanager
Energy Tracking	Energy Tracking Analytics	http://www.energytracking.com/ET_Analytics.htm
Energy WorkSite	Energy Expert	http://www.energyworksite.com/corporate/
EnergyPro	EnergyPro Solutions	http://www.energyprosolutions.com/about-energypro-solutions-ltd/
EnTechUSB	Interval Data Reporting & Monitoring Service	http //www.entech.co.uk/entechusb/Services.aspx?id=7
EnVINTA	ENTERPRIZE.EM	http //www.envinta.com/?page_id=65
eSight Energy Group	eSight	www.eSightEnergy.com
FirstEnergy Solutions—The E Group	Utility Risk Management	http://egrp.org/content/egroup/what_we_do.html
Good Steward Software LLC	EnergyCAP Enterprise	http://energycap.com/
Itron	Enterprise Energy Management Suite	https://www.itron.com/na/productsAndServices/Pages/Itron%20 Enterprise%20Edition%20Customer%20Care%20Suite%20%20CI.aspx
Kilojolts	EnerCop Suite	http://www.kilojolts.com/software/
LPB Energy Management	LPB's Utility Manager	http://www.lpbenergy.com/utility-manager.htm
MACH Energy	MACH Asset Manager	http://www.machenergy.com
McKinstry	EEM Suite Financial Management Solutions	http://www.mckinstryeem.com/

Appendix D. (Continued)

Vendor	Product Name	URL
Noveda Technologies	EnergyFlow Monitor	http://www.noveda.com/solutions/core-products/ energyflow-monitor/
Pace	EnControl Suite	http://www.paceglobal.com/EnergyCarbonManageme nt.aspx
Pulse Energy	Pulse Energy Management Software	http://www.pulseenergy.com/site/pulse-platform/
Schneider Electric	EPO Energy Profiler Online	http://www.powerlogic.com/product.cfm/c_id/2/sc_id/ 14/p_id/27#
Schneider Electric	EVO Energy View Online	http://www.powerlogic.com/product.cfm/c_id/2/sc_id/ 27/p_id/53#
Schneider Electric	PowerLogic ION EEM	http://www.powerlogic.com/product.cfm/c_id/2/sc_id/ 15/p_id/28
Utility Management Services	Utility Manager Pro	http://www.utilityaccounting.com/software_

Source: E Source. Last updated August 9, 2010.

APPENDIX E: FEDERAL GUIDANCE AND TOOLS FOR FACILITY MANAGEMENT, METERING, AND COMPLIANCE

E.1. Guidance for Federal Resource Management and Sustainability Goal Compliance

Guidance for Electric Metering in Federal Buildings (FEMP, 2006): Provides Federal building operators with interpretation and guidance related to the Energy Policy Act of 2005 (EPAct 2005) requirement for installing energy meters in all Federal buildings. An addendum to this guide addressing metering of natural gas and steam is under development. http://www1.eere.energy.gov/femp/pdfs/adv_metering.pdf

Facility Energy Management Guidelines and Criteria for Energy and Water Evaluations in Covered Facilities (FEMP, 2008): Contains guidelines for Federal agencies related to designating covered facilities, assigning energy managers, and performing comprehensive evaluations. http://www1.eere.energy.gov/femp/pdfs/eisa_s432_guidelines.pdf

DRAFT Guidance for the Implementation and Follow-up of Identified Energy and Water Efficiency Measures In Covered Facilities (FEMP, 2010):

Establishes steps for agencies to follow when implementing comprehensive evaluations, including how to perform a commissioning assessment, how to work with energy services companies (ESCOs) to install and monitor/verify retrofits, and how to use the Compliance Tracking System (CTS), an online system for reporting benchmarking and the status of evaluations to FEMP. Currently not available online.

Building Energy Use Benchmarking Guidance (FEMP, 2010): Interprets the requirement to benchmark the energy use of all "covered facilities" under EISA 2007, including the list of criteria for covered facilities and using a benchmarking system such as ENERGY STAR Portfolio Manager. http://www1.eere.energy.gov/femp/pdfs/eisa432_guidance

Energy Savings Assessment Training Manual (FEMP, 2010): Provides information on advancing self-sufficiency and using on-site energy audits. Currently not available online.

Federal Water Efficiency Best Management Practices (FEMP, 2010): Provides agencies with guidance to achieve water efficiency goals and assists with implementing Executive Orders 13123 and 13423. http://www1.eere.energy

E.2. General Guidance and Tools for Facility Management and Efficiency Projects

Building Upgrade Manual (EPA, 2008): Outlines a process for implementing strategic upgrades on commercial buildings, listing the most efficient measures and improvements in a practical order and providing case studies for specific building types. http://www.energystar.gov/index.cfm?c=business.bus_upgrade_manual

Continuous Commissioning^{SM} ***Guidebook for Federal Energy Managers*** (FEMP, 2002): Provides a detailed discussion of commissioning measures for Federal facilities, including HVAC, air handling units, water and steam systems, chillers and heat plants, and thermal storage, as well as how to monitor these measures over time. http://www1.eere.energy

Commissioning for Federal Facilities (FEMP 2008): Includes an overview of the general processes of commissioning new buildings as well as retrocommissioning, recommissioning, and continuous commissioning of existing buildings from a management perspective. http://www1.eere.energy

M&V Guidelines: Measurement and Verification for Federal Energy Projects, Version 3.0 (FEMP, 2008): Provides guidance on measurement and verification practices associated with Federal resource efficiency projects. Projects included cover areas such as energy efficiency and water conservation measures, construction, improved operations and maintenance, cogeneration, and renewable energy. http://www1.eere.energy.gov/femp/pdfs/mv_guidelines

Operations & Maintenance Best Practices: A Guide to Achieving Operational Efficiency, Release 3.0 (FEMP, 2010): Provides facility managers with a comprehensive overview of building operations and maintenance, including Computerized Maintenance Management Systems, commissioning, metering, and predictive maintenance. http://www1.eere.energy.gov/femp/pdfs/omguide_complete.pdf

Life-Cycle Costing Manual for the Federal Energy Management Program (NIST Handbook 135) (NIST, 1996): Explains, in detail, the principles of life-cycle cost analysis and integrates them with FEMP criteria on how agencies shall use life-cycle cost analysis in making decisions about investments in products, services, construction, and other projects. http://fire.nist.gov/bfrlpubs/build96/PDF/b96121.pdf

- ***Energy Price Indices and Discount Factors for Life-Cycle Cost Analysis —2010*** (NIST 2010): provides annual updates to the LCC Manual. http://www1.eere.energy
- ***Building Life-Cycle Cost*** software(BLCC5) provides computational support for analyzing capital investments in buildings using factors from the LCC Manual, including specific analysis capabilities for both appropriations-funded projects as well as energy savings performance contract (ESPC) or utility energy savings contract (UESC) projects financed with private capital. http://www1.eere.energy.gov/femp/information/cfm/register_blcc.cfm

Facility Energy Decisions System (FEDS) Model (PNNL 2010): Assists with optimizing life-cycle cost savings using a suite of measures that are selected based on user input about a building or group of buildings. A Windows-based software system, FEDS can be used for Federal facilities where multiple buildings use the same meter. http://www.pnl.gov/feds/

APPENDIX F: STATE OF FEDERAL SUSTAINABILITY INITIATIVES

Federal agencies are now required to address several overlapping and partially competing goals related to sustainability. Many of these goals are quantitative and require the use of specific tools or reporting systems to calculate and report results for compliance. While a number of tools are available for budgeting energy and emissions, four are most relevant to Federal sustainability goals: EPA's ENERGYSTAR Portfolio Manager, FEMP's Sustainability Report and Compliance Tracking System (CTS), and GSA's Carbon Footprint and Green Procurement Tool. Any of these tools can be used to calculate compliance with sustainability initiatives, as summarized in Table F1. (Note: GSA's Carbon Footprint and Green Procurement Tool is a bottom-up calculator aimed to assist agencies in greenhouse gas (GHG) calculations, which can then be reported in the FEMP Sustainability Report.)

Some of these requirements, such as GHG reporting under Executive Order 13514, are too new to have status updates as their first reporting has not yet occurred (EO 13514 reporting is scheduled for January 2011). However, both GHG requirements are tracked by FEMP, and preliminary data is available up to fiscal year 2009. Current compliance with these requirements is summarized in Table F2.

Table F1. Summary of recent Executive Orders and Statutes setting quantitative goals on agency facility energy and water use and GHG emissions

Order/Statute	Type of Sustainability Goal		
	Energy	Water	Greenhouse Gas
EO 13423	Decrease building energy use 3% annually or 30% by 2015, 2003 baseline	Decrease water use intensity 2% annually or 16% by 2015, 2007 baseline	
EO 13514		Decrease water use intensity 2% annually or 20% by 2020, 2007 baseline	Agency-wide, meet Scope 1 and 2 baseline and 2020 reduction goal by January 2010, Scope 3 by June 2010

Table F1. (Continued)

Order/Statute	Type of Sustainability Goal		
	Energy	Water	Greenhouse Gas
EPACT 2005	Implement building-level metering in all federal facilities Purchase renewable energy, 3% 2007-2009, 5% 2010-2012, 7.5% 2013 onward		
EISA 2007	Implement mandatory energy benchmarking Assign energy manager for covered facilities and institute energy and water audits every 4 years		

Represents requirements reported to FEMP Sustainability Report;

Represents requirements reported to EPA ENERGYSTAR Portfolio Manager;

Represents requirements reported to FEMP Compliance Tracking System.

Table F2. Current state of implementation of quantitative Federal sustainability goals

Order/Statute	Requirement	Quantitative Goal FY 2009	Current Status FY 2009 (All)	Current Status FY 2009 (Non-Defense)	Agency Range
EPAct 2005	Metering	N/A, 100% by 2012	80%	89%	30% – >100%[71]
EPAct 2005	Renewable Energy Purchase	3%	4.2%	4.8%	0.1% – 116%[72]
EISA	Facility Energy	12% reduction	13% reduction	18% reduction	0% – 36%
EISA	Facility Evaluations	50% (CV 2009)	38% - 45%	47% - 64%	0% – 100%
EO 13423	Facility Water	4% reduction	4.6% reduction	4.1% reduction	-15% – 22%

Perhaps most relevant to submetering, the Energy Policy Act of 2005 (EPAct 2005) requires Federal facilities to be metered for electricity whenever practical at the building scale by 2012. Already, 80 percent of Federal facilities and 89 percent of non-defense agencies are metered at the building scale as of fiscal year 2009. Agency compliance ranges from 30 percent to over 100 percent (representing agencies with buildings containing both standard and advanced meters).

In terms of facility energy use and facility water use, government-wide implementation met its goals for fiscal year 2009, with a 13 percent reduction in facility energy use over 2003 (compared to a goal of 12 percent), and a 4.6 percent reduction in facility water use over 2007 (compared to a goal of 4 percent). Because the DoD represents such a large share of overall facility resource use, the current status of Federal agencies other than DoD is shown separately. Overall, DoD has performed better than the average for water usage reductions partially due to prior metering requirements.[73] At the Federal agency level, results vary substantially, from a 0 to 36 percent reduction in energy use and a -15 to 22 percent reduction in water use, compared with the baseline year. It should be noted that the Federal Government is only meeting its goals for energy use reductions when taking source reduction credits for renewable energy purchases; without taking credit for the use, overall energy use reductions are only 9 percent compared to the goal of 12 percent.

Similarly, renewable energy purchase goals are on target, with a current proportion of renewable electricity of 4.2 percent compared to the goal of 3 percent for 2009. Agencies other than DoD are slightly higher at 4.8 percent. However, the goal will jump to 5 percent in fiscal years 2010 to 2012, and to 7.5 percent after that, so substantial investment in renewable energy generation or purchases will likely be needed in coming years.

In terms of comprehensive energy and water evaluations for the Energy Independence and Security Act (EISA) Section 432, the first round of evaluations is nearly halfway done, with most agencies evaluating around 50 percent of facilities as of November 2010. These evaluations are complicated by two factors, however—first, EISA evaluations are tallied on a calendar rather than fiscal year basis, and second, FEMP released three different metrics for evaluating buildings: total facility energy use, total number of facilities, and total facility square footage. As of FY 2009, preliminary data from FEMP's CTS shows that the evaluated share of facility energy use is 35 percent, though nearly 45 percent of facilities by number have been evaluated. From a square footage basis, 38 percent have been examined. By all three measures, agencies are making steady progress, approaching the calendar year target of 50 percent.

Similar to those objectives previously described, the results for evaluations by agency are extremely mixed, with some agencies (e.g., VA and USPS) evaluating a large fraction of their facilities (97 percent and 82 percent, respectively) and other agencies lagging significantly. An interesting trend relevant to metering is that there seems to be a negative correlation between average facility size and how quickly evaluations take place in different agencies. Audits at small facilities are less time-intensive and costly than those at large multi-building facilities such as military bases and laboratory campuses. Thus, facilities with the smallest average size, such as the USPS, have audited a large proportion of buildings, and facilities with the largest average size, such as the DoD, NASA, and DOE are affiliated with the lowest percentages of evaluated energy.

End Notes

[1] Vornado Realty Trust. http://www.vnooffice.com/sustainability

[2] DOE/EIA Annual Energy Outlook 2010, http://www.eia.doe.gov/oiaf/aeo/demand.html.

[3] Federal Energy Management Program, "Guidance for Electric Metering in Federal Buildings," February 3, 2006, http://www1.eere.energy

[4] Federal Energy Management Program, "Metering Best Practices: A Guide to Achieving Utility Resource Efficiency," October 2007, http://www1.eere.energy

[5] Wiley, J.A., J.D. Benefield, K.H. Johnson, "Green Design and the Market for Commercial Office Space," *J Real Estate Finan Econ* (2010) 41:228–243.

[6] Taken from: Mills, E., Mathew, P. "Monitoring-Based Commissioning: Benchmarking Analysis of 24 UC/CSU/IOU Projects," Report No. LBNL-1972E. Lawrence Berkeley National Laboratory, Berkeley, California, 2009.

[7] http://www.awwa.org/Resources/WaterLossControl.cfm?ItemNumber=48511

[8] U.S. Environmental Protection Agency, "State of the Art Submetering," ENERGY STAR Monthly Partner Web Conference, August 16, 2006. http://www.energystar.gov/ia/business /networking

[9] U.S. Environmental Protection Agency, "Best Management Practices for Colleges and Universities," 2007. http://www.epa.gov/region1/assistance/univ/pdfs/bmps/SCSU Submetering1-8-07.pdf.

[10] Grisham, A., Fleming, W.M., Long-Term Options for Municipal Water Conservation. *Journal AWWA*, 81-3, 34-42, 1989.

[11] Abrahamse, W., Steg, L., Vlek, C., & Rothengatter, T., A review of intervention studies aimed at household energy conservation. *Journal of Environmental Psychology*, 25, 273-291, 2005.

[12] NYSERDA, "Demonstration of New Submetering Technologies," Report 86-8, October 1986.

[13] NYSERDA, "Electricity Reduction in Master Metered Buildings Low Income Rate Impact Analysis," October 2009.

[14] Millstein, D., "Getting Green with Submeters," Electrical Contracting Products, July 2008.

[15] Vornado Realty Trust Energy Information Portal. http://www.vnooffice.com /sustainability

[16] The peak demand is lower as the probability that each individual meter registers a peak demand at identically the same instant is negligible.

[17] U.S. Environmental Protection Agency's ENERGY STAR®, "Sub-Metering Energy Use in Colleges and Universities: Incentives and Challenges," December 2002.

[18] Ibid.

[19] Life-Cycle Cost Analysis is codified in 10 Code of Federal Regulations Section 436 Subpart A—Methodology and Procedures for Life Cycle Cost Analyses. Life costing computational support is provided by the National Institute for Standards and Technology via the Federal Energy Management Program. ASTM Standard Practice E 917 provides additional guidance on performing an LCCA (ASTM International. "Standard Practice for Measuring Life-Cycle Costs of Buildings and Building Systems," E 917, *Annual Book of ASTM Standards: 2009*. Vol. 04.11. West Conshohocken, PA: ASTM International).

[20] ASTM International. "Standard Practice for Measuring Benefit-to-Cost and Savings-to-Investment Ratios for Investments in Buildings and Building Systems," E 964, *Annual Book of ASTM Standards: 2009*. Vol. 04.11. West Conshohocken, PA: ASTM International.

[21] ASTM International. "Standard Practice for Measuring Net Benefits and Net Savings for Investments in Buildings and Building Systems," E 1074, *Annual Book of ASTM Standards: 2009*. Vol. 04.11. West Conshohocken, PA: ASTM International.

[22] ASTM International. "Standard Practice for Measuring Internal Rate of Return and Adjusted Internal Rate of Return for Investments in Buildings and Building Systems," E 1057, *Annual Book of ASTM Standards: 2009*. Vol. 04.11. West Conshohocken, PA: ASTM International.

[23] NYSERDA, "Electricity Reduction in Master Metered Buildings Low Income Rate Impact Analysis," October 2009.

[24] Sullivan, G.P., R. Pugh, W.D. Hunt. "Metering Best Practices: A Guide to Achieving Utility Resource Efficiency," October 2007, http://www1.eere.energy

[25] Sullivan, G.P., R. Pugh, A.P. Melendez, W.D. Hunt. "Operations and Maintenance Best Practices," Release 3.0, August 2010 http://www1.eere.energy

[26] EPRI. 1996. *End-Use Performance Monitoring Handbook.* EPRI TR-106960, Electric Power Research Institute, Palo Alto, California.

[27] National Institute of Standards and Technology, "National Institute of Standards and Technology Policy on Traceability," http://ts.nist.gov/Traceability/Policy /nist_traceability_policy-external.cfm.

[28] Mane, L., "w – A Practical Approach," GE ESL Magazine, Summer 2005. http://www.geindustrial.com/Newsletter/fall05_submetering.pdf.

[29] Energy Information Administration, "Annual Energy Outlook 2009," page 119, table A5, http://www.eia.doe.gov/oiaf/archive/aeo09/pdf/0383(2009).pdf.

[30] Mills, E., "Building Commissioning: A Golden Opportunity for Reducing Energy Cost and Greenhouse Gas Emissions", Lawrence Berkeley National Laboratory, Berkeley, CA, July 2009. http://cx.lbl.gov/2009-assessment

[31] Water Use Case Study: Norwood Hospital, Massachusetts Water Resources Authority. http://www.mwra.state

[32] Ibid

[33] See http://www.bacnet.org/

[34] See http://www.echelon.com/communities

[35] DOD Unified Facilities Guide Specifications 25 10 10 and 23 09 23 use the LonTalk (CEA/ANSI 709.1b) protocol.

[36] Environmental Protection Agency, "ENERGY STAR Building Manual," Chapter 2, http://www.energystar.gov/ia/business/EPA

[37] Environmental Protection Agency, "ENERGY STAR® Performance Ratings Technical Methodology," http://www.energystar.gov/ia/business/evaluate_performance/General_Overview_tech_methodology

[38] U.S. Energy Information Administration, 2003 Commercial Buildings Energy Consumption Survey (CBECS), http://www.eia.doe.gov/emeu/cbecs/.

[39] Peterson, K., and H. Crowther. "Building EUIs." *High Performance Buildings* (Summer 2010): 40–49.

[40] Oak Ridge National Laboratory, "Benchmarking Building Energy Performance," http://eber.ed.ornl.gov/benchmark/homepage.htm.

[41] ENERGY STAR, "Portfolio Manager Overview," http://www.energystar.gov/index.cfm?c=evaluate_performance.bus_portfoliomanager_model_tech_desc.

[42] ENERGY STAR, "Guidelines for Energy Management Overview, Step 6.1: Measure Results," http://www.energystar.gov/index.cfm?c=guidelines

[43] Under the Information Technology Management Reform Act (Public Law 104-106), the Secretary of Commerce approves Federal Information Processing Standards that are developed by the National Institute of Standards and Technology (NIST) for Federal computer systems. In addition, NIST in accordance with its statutory responsibilities under the Federal Information Security Management Act (FISMA), Public Law 107-347, is responsible for developing information security standards and guidelines, including minimum requirements for federal information systems. For more information, see http://csrc.nist.gov. The National Security Agency (NSA) develops guidance for DoD's national security information and information systems, in accordance with National Security Directive 42. See also DOD Unified Facilities Guide Specifications 25 10 10 and 23 09 23 focused on control protocols.

[44] Data specifications based upon work sponsored by the California Energy Commission Public Interest Energy Research Program (PIER) and the Building Technologies Program of the U.S. Department of Energy, http://cbs.lbl.gov/performance-monitoring/specifications/.

[45] "Mining Your Meter for Savings." Platts, a Division of the McGraw Hill Companies, Inc., 2004. http://www.oldsanteecanalpark.org/portal/page/portal/santeecooper/mybusiness/energymanagementtips/ miningyourmeterforsavings.pdf.

[46] Lawrence Berkeley National Laboratory, "Building Energy Information Systems and Performance Monitoring (EIS-PM)," http://eis.lbl.gov/.

[47] "Visualizing Building Information: Using information feedback to educate and influence building managers and occupants." *Centerline* (Winter 2009), 3–9, http://www.cbe.berkeley.edu/centerline/winter2009.pdf.

[48] Ibid.

[49] Granderson, J., M.A. Piette, G. Ghatikar, P. Price. "Building Energy Information Systems: State of the Technology and User Case Studies," Report No. LBNL-2899E. Lawrence Berkeley National Laboratory, Berkeley, California, 2009.

[50] Ibid.

[51] Vornado Realty Trust, undated memorandum, http://www.vnooffice.com/sustainability

[52] Federal Energy Management Program, "Guidance for Electric Metering in Federal Buildings," February 3, 2006, http://www1.eere.energy

[54] Federal Energy Management Program, "Metering Best Practices: A Guide to Achieving Utility Resource Efficiency," October 2007, http://www1.eere.energy

[54] Office of the President, Executive Order 13514, "Federal Leadership in Environmental, Energy, and Economic Performance," *Federal Register* 74, no. 194 (8 October 2009), http://www1.eere.energy

[55] "Comparison of Commercial Building Energy Rating and Disclosure Mandates," Institute for Market Transformation, 2010, http://imt.org/files/FileUpload/files/Benchmark/BenchmarkingComparisonMatrixLatest.pdf.

[56] New York City Council Initiative 476-A, 2009, http://www.nyc.gov/html/planyc2030/downloads/pdf/476.pdf.

[57] Austin City Council Energy Conservation Audit and Disclosure Ordinance, 2008, http://www.austinenergy.com/About%20Us/Environmental%20Initiatives/ordinance/index.htm.

[58] Seattle City Council Ordinance 123226, 2010, http://clerk.ci.seattle.wa.us/~scripts.exe?d=CBOR&s1=116731.cbn.&Sect6=HITOFF&l=20&p=1&u=/~public/cbor2.htm&r=1&f=G.

[59] District of Columbia Council Clean and Affordable Energy Act of 2008, http://www.agobservatory.org/library.cfm?refID=106301

[60] State and Local Governments Leveraging Energy Star, http://www.energystar.gov/ia/business/government/State_Local_Govts_Leveraging_ES.pdf.

[61] ENERGY STAR®, "Guidelines for Energy Management Overview," http://www.energystar.gov/index.cfm?c=guidelines.guidelines_index.

[62] U.S. Energy Information Administration, 2003 Commercial Buildings Energy Consumption Survey (CBECS), http://www.eia.doe.gov/emeu/cbecs/.

[63] ASHRAE Building Energy Quotient Program, http://buildingeq.com/.

[64] The Chicago Energy Code is viewable as Chapter 18 of the Chicago Municipal Code at http://www.amlegal.com/ nxt/gateway.dll/Illinois/chicago_il/municipalcodeofchicago?f=templates$fn=default.htm$3.0$vid=amlegal:chicago_il. See section 18-13-505.7 for electrical energy consumption for buildings having individual dwelling units.

[65] ANSI/ASHRAE/USGBC/IES Standard 189.1-2009, Standard for the Design of High-Performance Green Buildings Except Low-Rise Residential Buildings.

[66] ICC International Green Construction Code. Public Version 1, available for download via links at http://www.iccsafe.org/cs/igcc/pages/default.aspx.

[67] Abrahamse, W., L. Steg, C. Vlek, and T. Rothengatter, "A review of intervention studies aimed at household energy conservation," *Journal of Environmental Psychology* 25, no. 3 (2005): 273–291; Darby, S, "The effectiveness of feedback on energy consumption," A Review for DEFRA of the Literature on Metering, Billing and Direct Displays, 2006; Faruqui, A., S. Sergici, and A. Sharif, "The impact of informational feedback on energy consumption—A survey of the experimental evidence," *Energy* 35, no. 4 (2010): 1598–1608; Roberts, S. and W. Baker, "Towards effective energy information: Improving consumer feedback on energy consumption," Bristol, UK: Center for Sustainable Energy, 2003.

[68] Ibid

[69] Henryson, J., T. Hakansson, and J. Pyrko. "Efficiency in buildings through information—Swedish perspective." *Energy Policy* 28, no. 3 (2000), 169–180.

[70] Craig, C. S., and J. M. McCann. "Assessing Communication Effects on Energy Conservation." *The Journal of Consumer Research* 5, no. 2 (1978): 82–88.

[71] FEMP statistics track total number of meters: buildings may have both a smart meter and a standard meter, hence values may exceed 100%.

[72] EPA generates and purchases more renewable energy than the total amount it consumes.

[73] DoD Instruction 4170.11, Installation Energy Management first released on November 22, 2005 and revised on December 11, 2009 requires metering electricity, natural gas, steam, and water at appropriate facilities for all new military construction, major renovations, and ESPC projects.

INDEX

A

B

D

E

N

O

P

S

T

U

V

W

Y